Sustainable Forest Management

Sustainable Forest Management

Edited by **Aduardo Hapke**

New York

Published by Callisto Reference,
106 Park Avenue, Suite 200,
New York, NY 10016, USA
www.callistoreference.com

Sustainable Forest Management
Edited by Aduardo Hapke

International Standard Book Number: 978-1-63239-582-5 (Hardback)

Printed in the United States of America.

Contents

Preface

Sustainable forest management (SFM) is an old concept. But its popularity has grown in the last few decades due to the public concern regarding the significant decline in forest resources. The implementation of SFM is usually achieved with the help of criteria and indicators (C&I). A number of countries have developed their own sets of C&Is. The book covers some of the current researches carried out to test the recent indicators, to look for new indicators and to establish new tools. It contains original research studies on socio-economic functions and decision making tools. These studies exhibit the recent researches done to provide forest managers with effective tools for choosing between distinct management strategies of improving indicators of SFM.

The information contained in this book is the result of intensive hard work done by researchers in this field. All due efforts have been made to make this book serve as a complete guiding source for students and researchers. The topics in this book have been comprehensively explained to help readers understand the growing trends in the field.

I would like to thank the entire group of writers who made sincere efforts in this book and my family who supported me in my efforts of working on this book. I take this opportunity to thank all those who have been a guiding force throughout my life.

Editor

Section 1

Socioeconomic Functions

Economic Valuation of Watershed Services for Sustainable Forest Management: Insights from Mexico

G. Perez-Verdin[1,*], J.J. Navar-Chaidez[1], Y-S. Kim[2] and R. Silva-Flores[3]
[1]*Instituto Politécnico Nacional*
[2]*Northern Arizona University*
[3]*Private Consultant*
[1,3]*Mexico*
[2]*United States of America*

1. Introduction

Ecosystem services are the benefits that people obtain from ecosystems (Brauman et al., 2007). Recognizing the importance of the services provided by ecosystems for human well-being is not a new idea, going as far as Plato (Feen, 1996) and the economic conceptualization of ecosystem values (Coase, 1960; Feen, 1996). However, the scientific and practical interests in assessing and trading ecosystem services have not gained momentum until the 1990s when pioneering works by Daily (1997) and Costanza et al. (1997) galvanized the field. Among the ecosystem services that received increasing attention in the recent years are the hydrological services due to the role of water as a vital, and sometimes decisive, element in human life (Pare et al., 2008). Hydrologic services encompass a range of benefits that terrestrial ecosystem produces in terms of freshwater. These services can be grouped as: improvement of extractive water supply, improvement of in-stream water supply, water damage mitigation, provision of water related cultural services, and water-associated supporting services (Brauman et al., 2007).

The majority of hydrological services take place in the highlands of forest watersheds (Messerli et al., 2004). In these areas, upland forest watersheds work as a source that collects, manufactures, and distributes water and provides hydrological services to lowlands (Neary et al., 2009). Various components of the water cycle (i.e., evaporation, infiltration, surface run-off) critically depend on forest cover. If the forest cover is affected, so it will be the quality and quantity of the water provided to downstream users (Brown et al., 2005). In developing countries, such as Mexico, changes in forest cover are caused among other things by the local economic conditions in which landowners live. While searching for basic needs (food and shelter), they exercise excessive pressure over the forests eventually triggering forest fragmentation and deforestation (Perez-Verdin et al., 2009).

Based on the methods used for their economic valuation, hydrological services can be classified into two broad categories of values: marketed and non-marketed. The economic

---

* Corresponding Author

value of the former is reflected through the market price determined mainly by its demand and supply (i.e., drinking water) while the latter, traded under imperfect markets, requires a more complex evaluation that involves evaluating consumer's preferences and behavior (i.e., evaluation of recreation sites). The sum of these services gives the total economic value (TEV) of a forest watershed. Because of the quasi-public good nature of hydrological services and the presence of externalities, failure to recognize the TEV of a watershed can lead to depletion, degradation, and overexploitation of forest resources and eventually loss of social welfare (Plottu & Plottu, 2007).

Recently, research has focused on assigning economic values to environmental services to redirect policies for sustainable forest management. The intention is to help landowners reduce the impact of externalities by giving monetary incentives and implement best management practices to regulate the quality/quantity of water (Pagiola et al., 2003; Muñoz-Piña et al., 2008). Among the new schemes include the formal articulation of incentive-based instruments, such as Payments for Ecosystem Services (PES) and Markets for Ecosystem Services (MES) (Jack et al., 2008; Gómez-Boggerthun et al., 2010). While the design and operation of various international PES and MES programs have been started by local governments, many of them now promote the participation of the private sector, non-government organizations, and the general public (Paré et al., 2008).

The major objective of this chapter is to underline the importance of assigning economic values to hydrological services as a means to achieve sustainable forest management. The paper first introduces critical inputs of the water balance and best management practices for watershed resources. It also describes the types of watershed services and how they can be valued. The paper then analyzes the cases where non-market valuation techniques have been implemented for various types of watershed services in Mexico. And finally, it discusses the operation of a Mexican PES program and its impact on watershed services.

2. Water balance and best management practices

The assessment of available water resources is central to economic valuation of hydrological services. The economic valuation of water resources involves knowledge of the supply and demand sides and eventually to the search for effective management policies. The determination of available water within a watershed is given by the water balance and depends on the magnitude of inputs and outputs and the storage capacity. The basic input is precipitation (P_T) and is either lost to evaporation (E_V) and transpiration (T_R) or routed through small pathways of overflow and interflow to give surface runoff (Q) and infiltration (I) (Hiscock, 2005). Thus, the water balance model, estimated for a given period of time ($\partial A/\partial t$), is the difference between inputs and outputs. The larger the difference between inputs and outputs, the more supply water there is to end users. In this case, Inputs= P_T and Outputs = $I + E_V + T_R + Q$. Therefore, the water balance can be expressed as:

$$\partial A/\partial t = P_T - (I + E_V + T_R + Q) \tag{1}$$

In mountainous forest watersheds, precipitation is partitioned into throughfall, interception loss, and stemflow (Navar, 2011). Throughfall is the rainfall portion that reaches the ground by passing directly through or dripping from tree canopies. Interception loss is the rainfall retained on the canopy that evaporates back to the atmosphere; it is composed mainly on the amount of precipitation stored by canopies and the evaporation of stored canopy water.

Stemflow is the rainfall portion that flows to the ground via trunks or stems (Dunkerley, 2008). Litter retains part of the throughfall and stemflow and infiltrate into the mineral soil increasing soil moisture content. Evapotranspiration is the amount of water vapor that leaves soil and vegetation via evaporation and transpiration. Factors that control evaporation from soils are the current water content, the water content at wilting point, and the soil water content at field capacity. Factors that affect transpiration are the type of vegetation, density, and age.

Conventional forest management practices, that include logging and grazing, affect tree density, canopy cover, and tree composition and structure (Brown et al., 2005). Hydrologic studies in the United States have demonstrated that selective harvesting and clear-cutting promotes increased discharge because of a reduction of stand density and canopy cover that demand less water for transpiration (Swank et al., 1988; McBroom et al., 2008). Non-conventional forest disturbances that cause tree mortality include forests fires, pests and diseases, strong winds, etc. Forest fires of large spatial scales and severity, in addition to tree mortality, also cause soil water repellency (Martin & Moody, 2001). Water repellency reduces infiltration and often promotes surface runoff and soil erosion beyond any other forest disturbance (Pierson et al., 2008). In general, tree mortality beyond natural causes reduce interception loss and transpiration leaving more net precipitation (throughfall) for other processes such as soil moisture content, aquifer recharge, and surface runoff (Brown et al., 2005; Ikawa et al., 2009). In addition, streamflow and aquifers are enriched with sediments and chemicals washed out from the soil that reduces usability. Other human-related disturbances are road construction and maintenance, and harvest-related activities that promote soil compaction and reduce soil infiltration at specific places in the watershed.

The aim of best management practices (BMP) is to reduce the effect of non-point and point sources of degradation that affect water quality and quantity (McBroom et al., 2008). Examples of non-point sources, which are characterized by a widespread and diffused generation, include cropland, harvesting areas, animal feedlots and grazing lands, impervious surfaces (e.g., roads, land rocks, deforested sites, urban areas), and construction sites (Neary et al., 2009). Transport of sediments, organic matter, and nutrients, such as nitrogen and phosphorus are examples of point sources. Harvesting, grazing, and agriculture can lead to increased rates of runoff and erosion. Rates of material export from impacted watersheds to water resources, while highly variable within and between land uses, exceed those for natural or undisturbed land uses (Andreassian, 2004). Because of this characteristic, the application of BMP is mainly oriented to reduce the effect of non-point sources.

Effective BMPs to reduce the effect of non-point source loads should target changes in current land-use practices, construction and operation of equipment, machinery, and the use of structures to retain or otherwise control the movement of water and material (McBroom et al., 2008; Neary et al., 2009). Also, effective BMPs need to consider the local conditions (e.g., geology and soils, topography, climate, and hydrology), landowner expectations, and the nature of the source of the polluting material (e.g., harvesting, grazing, or agricultural land uses) in which impacts are occurring. Overall, watershed BMPs are oriented to (1) minimize soil compaction and bare ground coverage, (2) separate exposed bare ground from surface waters, (3) exclude fertilizer and herbicide applications from surface waters, (4) inhibit hydraulic connections between bare ground and surface waters, (5) avoid disturbance in steep convergent areas, (6) provide a forested buffer around streams, and (7) build stable road surfaces and stream crossings (Jackson & Miwa, 2007; Neary et al., 2009).

In Mexico, the national water, environmental protection, and forest laws are the basis for regulating watershed management practices. Coupled with the federal laws, almost every state in the country has specific regulations that complement those issues where the federal laws do not apply. Based on this set of laws and regulations, common examples of BMPs that involve forest vegetation and water include: the provision of forested buffer around streams, stabilization and closure of third-order roads immediately after harvesting, construction of culverts on primary and secondary roads crossing streams, pre-harvest planning for cutting, skidding and loading zones to avoid increasing hydrologic and sediment source connectivity to stream channels, and the perpendicular arrangement of forest residues to reduce soil erosion, among others.

In the past, the implementation of these BMPs was adopted by landowners who would evaluate the cost and benefits in either doing another activity or doing nothing. Since these practices, which we have identified as externalities, would reduce their economic profits, many landowners did not comply with the regulations leading to increased rates of erosion and sedimentation (Muñoz-Piña et al., 2008). Nowadays, the cost of BMPs is mostly shared with the government; however, the private sector, non-government organizations, and the general public are participating as well. This type of cost-share programs, which embrace the known concept of internalizing externalities, is discussed in section 4 of this chapter.

3. Economic valuation of watershed services

The need of economic valuation of watershed services stems from their quasi-public and non-rivalry nature, the presence of externalities, and scales of production (Pattanayak, 2004; Brauman et al., 2007; Plottu & Plottu, 2007). In a market economy, watershed services without economic values will not be provided at optimal levels. The quasi-public, non-rivalry nature implies that it is difficult, if not impossible, to exclude an individual from using watershed services (e.g. soil retention), and several individuals can use them simultaneously without diminishing each other's use values. The presence of externalities means that the economic benefits of users of these services will not be deviated to compensate providers. And regarding the scale of production, these services are characterized by economies of scale in production; the larger the watershed, the lower the marginal costs (Pattanayak, 2004).

Valuation of watershed services also implies understanding the different types of benefits a watershed offers to ecosystems and society. A forest watershed not only functions like a basin which receives and stores water from precipitation, surface runoff, or infiltration, but also cleans water, retains sediments, provides habitats for wildlife, sinks CO_2, and offers many environmental amenities for humans (Brauman et al., 2007; Locatelli & Vignola, 2009). Some of these benefits can be valued through conventional methods that use market-based approaches. For example, the useful life of a dam can be valued through estimations of the rate of sedimentation and the years left to sustain fish. Other benefits require detailed information and more complex approaches that estimate for example the value of environmental services for present, future generations, or consider the presence of externalities (Field, 2008). For example, if fewer recreation opportunities are provided in the watershed, due to water loss resulting from harvesting or grazing, recreationists may act and eventually offer a fee to preserve the watershed and recover the loss of recreation opportunities. In this section, we provide a brief summary of the different watershed values and the means to estimate them.

3.1 Watershed values

For the purpose of this work, we will focus on two main types of watershed values: use and non-use values (Freeman, 2003; Field, 2008). Use values, which consist of consumptive and non-consumptive uses, refer to the situations where people directly or indirectly interact with resource use (Field, 2008). Consumptive use values are derived from extractive resource uses such as timber, commercial fishing and hunting, and the use of water for irrigation and drinking. Examples of non-consumptive uses values are benefits from resources with a minimal or imperceptible extraction and include those from boating, swimming, ecotourism, and camping.

Non-use or passive-use values refer to the situations in which people place monetary values on resources independent of their present or future use (Field, 2008). For example, people may be willing to support a long-term program intended to maximize water quality even though their offspring, not they, will receive the benefits. Despite the controversy that these types of values should not be considered in mainstream economics, because they reflect altruism and difficulty to assess, Freeman (2003) argues that non-use values can be defined within a utility theoretical framework and should be considered as public goods. Freeman further contends that ignoring non-use values could lead to wrong policies and resource misallocation.

The rationale for assigning values to watershed services also lies on the many biochemical cycles that take place in the watershed, the water and soil conservation functions, and the provision of wildlife habitats and amenities (Pearce, 2001; Pattanayak, 2004; Brauman et al., 2007). Water is the principal medium in which many chemical reactions occur and watersheds provide a variety of conditions in which those chemical reactions take place (Ward & Trimble, 2004). Water, Carbon, Nitrogen, Oxygen are among the key elements whose maintenance depends on the management of forest watersheds. Altering these cycles could interrupt the flow of environmental services, particularly water, to downstream communities (Figure 1). Therefore, the main question is how these hydrological processes, defined by a local drainage unit, can be manipulated to be fairly useful to society.

Figure 1 shows the relationship between hydrological processes and economic values to humans. A change in physical or chemical properties of water causes a change in the quality and quantity of the liquid provided. Discharges from non-point pollution sources affect the quality of water and force resource managers to use expensive processes, equipment to clean the water. Conversely, to address the feedback loop, excessive fishing may cause a change in the fish population. Estimating an improvement of watershed benefits involves the use of economic models to determine the monetary units people place on both use and non-use values (Freeman, 2003).

The TEV is a concept that illustrates the whole worth of ecosystem services. Due to the nature of some services, hypothetical markets are created to elicit values through a variety of economic techniques, including: (a) direct market valuation approaches, (b) revealed preference approaches, and (c) stated preferences approaches (Freeman, 2003; Champ et al., 2003;). Direct market valuation methods use data from actual markets and thus reflect actual preferences or costs to individuals. Revealed preference techniques are based on the observation of individual choices in existing markets that are related to the ecosystem service subjected to valuation. Stated preference approaches simulate a demand for ecosystem services by means of surveys on hypothetical changes in the provision of ecosystem services (TEEB, 2010). Selection of the best technique depends on the objectives of the researcher, the type of use values, and the type of ecosystem services under evaluation.

Again, due to the nature of some watershed services, uncertainty is an issue that must be considered in every valuation work. As suggested by TEEB (2010), one way to deal with uncertainty is the use of the data enrichment or data fusion approach which combines the use of revealed and stated preference methods. The main advantage of these hybridized approaches is that they overcome technical uncertainty due to application of valuations tools and uncertainty with regard to preferences about ecosystem services. However, their application generally depends on available financial, human, or time resources.

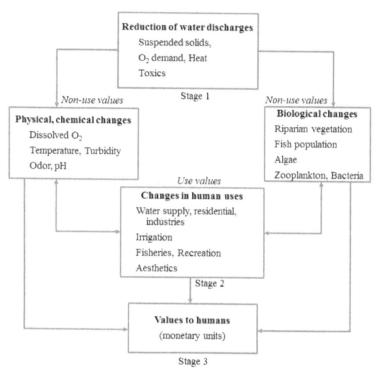

Fig. 1. Types of hydrological values and flow of services to society. The sum of use and non-use values gives the total economic value (From Freeman, 2003, page 31).

3.2 Theoretical framework of economic valuation

Because of the diversity of watershed benefits, which include use and non-use values, placing monetary units depends on the type of services provided, the actual and desired conditions of the watershed, and people's social status (Freeman, 2003; Brauman et al., 2007). Although valuation of all watershed benefits is possible, many studies focus on few or single services. The most common benefits include drinking, irrigation, wildlife habitats, prevention of soil erosion, flood protection, fisheries, and hydropower (see Pearce, 2001; Locatelli & Vignola, 2007, for a literature review of watershed services). To account for reliable estimations of the watershed value, information on the extent of the change in quality and/or quantity of the service is required. The marginal value, the extra monetary units a person would be willing to pay for an additional unit of the service, depends on the magnitude of the change a person expects with her/his contribution as well as on the beginning and ending points of that change (Brauman et al., 2007).

Concerning to watershed services, willingness-to-pay (WTP) is the maximum amount of income an individual will pay for an improvement in current conditions of the watershed, or the maximum amount of money to avoid a decline in those current conditions (Freeman, 2003). The WTP measure for valuing watershed services is a function of a vector of individual´s social characteristics (such as income, education, family size, among others), the price (p), and quantity of the service (q) (Freeman, 2003). Theoretically, WTP can be expressed as either in terms of an utility function $V(p, q, y)$:

$$V(p, q^*, y - WTP) = V(p, q, y) \tag{2}$$

or in terms of the minimum expenditure function $m(p, q, u)$,

$$WTP = m(p, q, u) - m(p, q^*, u), \text{ when } u = V(p, q, y) \tag{3}$$

where y is income and q^* represents a new condition or improvement in the watershed service $(q^* > q)$. The WTP is thus the amount of money to pay that would make such individual indifferent between the current condition (y and q) and the new, improved state $[(y-WTP), q^*]$.

To estimate the economic value of watershed services, particularly non-consumptive or non-use values, typically researchers use a stated preference technique called Contingent Valuation (CV). This technique employs survey-based information to directly elicit households' preferences and build a contingent market through which respondents may state their willingness to pay for a specified provision change in a particular service (Mitchell and Carson 1989). The CV approach first involves describing the current situation of a non-market good, how it can be improved, and then asking respondents whether or not they would pay for the improvement of the good (Boyle 2003). It is called contingent valuation, because people are asked to state their willingness to pay contingent on a specific hypothetical scenario and description of the environmental service (Carson & Groves 2007). The willingness-to-pay results can then be used by decision makers to weigh policy options. Details on CV description can be found in Mitchell & Carson (1989), Boyle (2003), Schlapfer (2008), TEEB (2019), among others.

4. Valuation of watershed services in Mexico

In recent years, various studies have been conducted to estimate the value of watershed services using non-market valuation techniques in Mexico. To document these cases, several sources of information where a consistent valuation approach was used were reviewed in this chapter. The first information source included a literature search from all available databases (e.g. Web of Science) and the web for nonmarket valuation studies. A brief review of the abstracts and introductions served to select articles directly related to watershed services and the valuation approach. Second, all articles relating to the topic were thoroughly reviewed to identify the main watershed services and other information needed to be considered. The search also included the citations of published articles to find any unpublished data or papers. Besides the WTP amount and the watershed service being evaluated, additional information collected in the review was altitude, latitude, longitude, and precipitation. The search eventually gave 13 cases including Mexico City and other cities located across the country. The watershed services ranged from wildlife habitat preservation, soil retention, and recreation, to drinking, irrigation, fishing, and hunting. The cases identified were compiled and georeferenced in a geographical information system (GIS) database.

Table 1 shows the cases included in the literature review. Most of the studies were located in high elevation areas (e.g., more than 1,000 meters above sea level) which gave indication of the relevance of the watershed highlands to provide environmental services, and the need to protect them. The WTP, obtained through the contingent valuation approach, ranged from US$ 0.45 to 15.8 per month and household, being the Mexico City the case with the highest WTP. These figures represent between 0.33 and 11.8% of the 2011 per-capita minimum wage (the minimum wage is US $134/person/month; DOF, 2010). The main types of services provided by the watersheds were wildlife habitat, drinking, and soil retention. The most common management practices proposed in the studies were reforestation, soil conservation works, and reducing harvesting, grazing and risk of fire, among others.

It is important to note that in many studies it was difficult to clearly identify the main watershed service. During the search, several works were discarded due to the inconsistency of valuation approaches, the service being evaluated, and the type of WTP units (for example, WTP was expressed in $/month/person, $/year/household, $/visit, etc). Out of the 25 studies reviewed, only those listed in Table 1 were selected since they allowed cross-site comparisons. Based on the predominant service, each case was classified into two major groups: those with consumptive use values (e.g., drinking, fishing, irrigation and hunting) and those with non-consumptive use values (ecotourism, wildlife habitat, recreation, soil retention); the latter also included non-use values. The classification yielded seven cases in the first group and six in the second. To test for WTP differences in the type of use values, one-way analysis of variance indicated that there was no significant relationship in the WTP[†] ($n=13$, $F=2.541$, $p=0.14$). Neither there was for elevation ($n=13$, $F=0.001$, $p=0.99$) and moisture index ($n=13$, $F=0.978$, $p=0.34$), the two additional physical variables of the cities. The lack of significance in the WTP differences means that individuals appreciate both consumptive and non-consumptive uses similarly. However, in practical terms, the individual benefits estimated for consumptive use values were 47% higher than those for non-consumptive use cases.

4.1 Government-supported watershed markets

Various Latin-American countries have started programs to intensify the production of watershed services in forest ecosystems. In 2003, Mexico launched an innovative PES program to help landowners to protect forest watersheds in critical areas of the country. The program, called in Spanish as *Pago de Servicios Ambientales Hidrológicos* (PSAH), had three main goals: to reduce deforestation in areas with severe water problems, apply best management practices for sustainable forestry, and reduce illegal logging (Muñoz-Piña et al., 2008). The PSAH consisted of direct payments to landowners, whose lands were mostly covered by temperate or tropical forests, during a 5-year period in which landowners executed a series of BMPs to protect the watersheds. Part of the PSAH's innovative approach is that it was funded through an earmarked portion of federal fiscal revenues from water

---

† Due to the small sample size, differences between the use values were also evaluated with the non-parametric Mann-Withney test. Results corroborated the results of no significant differences for WTP, elevation, and moisture.

fees, so the program involved users and producers of environmental services. The payment, offered as an economic compensation or subsidy, was based on the opportunity cost of using the land for agriculture or livestock (Muñoz-Piña et al., 2008), not on the non-market valuations we have discussed above. Initially, it oscillated between US$ 23 and 30 per hectare depending of the type of forest (CONAFOR 2004)‡.

As expected, the PSAH received various criticisms. The government used the opportunity costs of the two primary economic activities (agriculture and livestock) to estimate the compensation. Though there are no official reports, this was probably due to the type of information available initially. Government officials have said that these payments are currently under evaluation and will be reassessed with new information based on market and non-market methods. Also, the PSAH has been regulated by the government itself who

Study site	Watershed service [a]	Type of use value [b]	Elevation (meters)	Moisture index [c]	Adjusted WTP (US$/month) [d]	Source
Ciudad Obregon, SON	WH, F, SBR	NC	35	0.146	6.12	Ojeda, et al. (2008)
San Luis Rio Colorado, SON	F,H, SBR	C	40	0.055	6.39	Sanjurjo (2006)
Parral, CHIH	D	C	1,620	0.089	8.91	Vasquez et al. (2009)
El Salto, DGO	D,SR	C	2,540	0.250	2.08	Silva-Flores et al. (2010)
Tapalpa, JAL	I,D	C	1,950	0.135	9.10	Lopez-Paniagua, et al. (2007)
Mexico City, DF	D	C	2,240	0.064	15.81	Soto and Bateman (2006)
San Cristobal de las Casas, CHIS	D,WH	C	2,120	0.306	1.82	Gutierrez-Villalpando (2006)
Tepetlaoxtoc, EDOMEX	WH	NC	2,300	0.088	4.98	Jimenez-Moreno (2004)
Oaxaca, OAX	WH	NC	1,555	0.105	3.11	Garcia-Angeles (2006)
Tlaxco, TLAX	WH, SR	NC	2,588	0.074	1.83	Orozco-Paredes (2006)
Metztitlan, HGO	WH, SR	NC	2,080	0.091	0.45	Monroy-Hernandez (2008)
Alamos, SON	WH, SR, D	NC	400	0.046	8.23	Chan-Yam (2007)
La Paz, BCS	D	C	10	0.048	10.15	Aviles-Polanco, et al. (2010)

[a] WH, Wildlife habitat; D: Drinking; I, Irrigation; F, Fishing; H, Hunting; SBR, Scenic Beauty and Recreation; SR, Soil Retention
[b] C= Consumptive, NC = Non-consumptive
[c] Based on precipitation and evaporation data (Willmott & Feddema, 1992). The moisture index goes from 0 to 1, where dryer areas tend to zero.
[d] Based on Februrary-2011 price levels (US$1 = MEX$13) and 10-year average of the National Consumer Price Index =4.03%)

Table 1. Willingness-to-pay for watershed services in Mexico

‡ We tried to compare the PSAH payments to those WTP values extracted from literature (See Table 1). The comparison turned difficult due to the differences in methods, sampling issues, and monetary units.

acted as a monopsonistic buyer on behalf of water users (Muñoz-Piña et al., 2008). The government basically established a price and waited for landowners to offer their forests for conservation. In retrospective, some landowners may have rejected the program because the compensation was not enough to fully cover transaction and opportunity costs. In addition, the initiators of the program never considered a baseline to monitor the impacts of the economic compensations on the quality/quantity of water. Today, evaluating the performance of the first periods of the program is difficult due to the lack of a monitoring plan (Consejo Civil Mexicano, 2008).

Despite of these and other criticisms, the PSAH has endured and contributed to sustainable forest management by offering landowners more incentives to provide environmental services, while clarifying and defining property rights, thus reducing the impact of externalities (Muñoz-Piña et al., 2008). In the first years of operation, the program had paid almost US $200 million and protected about 1.5 million has of strategic watersheds (Chagoya & Iglesias, 2009). The PSAH also has received full support from the Mexican Congress which recently authorized the participation of state and local governments, non-government organizations, private entrepreneurs, and society to increase the funds. Examples of this type of mixed funds are found in Centro Montaña de Guerrero; Tehuacan, Puebla; Coatepec and Texizapa in Veracruz; Cupatitzio, Michoacan; and Chinantla Alta, Oaxaca; among others (Paré et al., 2008). Most of the collected fees have been used to implement selected best management practices in the watersheds' highlands.

The examples of multi-stake voluntary participation in the payment for environmental hydrological services have received ample attention due to the commitment of the multiple parties to promote sustainable forest management. Although programs like PSAH are not in themselves sufficient conditions for sustainable forest management, they are necessary conditions for efficient policy making. Assigning property rights to providers and consumers help delineate the responsibilities of each group. The former receives an economic compensation to reduce the effect of externalities in the management of forest resources. The latter express their demand for environmental services through their WTP for receiving a better quality watershed service. The interaction between providers and consumers helps partially correct market failures and eventually reduce forest degradation. Programs like PSAH not only generate the necessary funds for forest conservation, but also will increase the quantity/quality of watershed services (Pagiola, et al., 2003). The future of PSAH and similar programs lies in the clear definition of the real value of watershed services, correct assignment of property rights, and the continuity of funds.

5. Conclusions

This chapter discussed the relevance of valuing watershed services to achieve a sustainable management of forest resources in Mexico. It presented a simple method to estimate water balance and identified BMPs, discussed the main types of values a watershed can offer, how they can be valued, and examples of cases based on non-market valuation and government-supported programs. Due to their non-exclusive, non-rival characteristic, watershed services need to be economically valued using diverse approaches to be produced at optimal levels. Their valuation through opportunity costs may not reflect the total economic value, particularly of non-use values.

The Mexican PSAH, one of the largest in its type, is a clear example of the international concern for redesigning effective management policies for watershed resources (Muñoz-Piña et al., 2008). However, there are still a number of challenges for mainstreaming this type of programs. Turner & Daily (2008) summarized three key constraints that need to be overcome before ecosystem services become operational: 1) information failure, where decision-makers lack scale-relevant detailed information on important ecosystem services and their tradeoffs; 2) institutional failure, where property rights and institutions are lacking to ensure legitimacy and equity; 3) market failure, where investments in long-term ecosystem health can be discouraged due to shared benefits and missing prices for public goods.

We have reviewed several cases of non-market valuation that estimated the benefits of watershed services in Mexico (Table 1). Results indicate that there is no significant relationship between WTP, moisture index, and elevation with the two types of values (i.e., consumptive and non-consumptive uses). Considering the low number of cases found in the literature, more research is clearly needed to evaluate the relationship between WTP and the benefits of environmental services, and motivate the interest for creating markets, particularly of non-use values. Here, researchers must incorporate a diversity of geographical areas and services to scale up these markets and incentive programs. They also must employ appropriate valuation tools to tackle the problems associated with reliability of results such as survey design, definition of contingent valuation scenarios, and testing for survey variations of results (Wittington, 2002). More work is necessary to understand the benefits of use and non-use values of watershed services, disseminate the results of pilot projects (success stories), and incorporate all interested sectors of society. This kind of work would increase public's level of awareness and their perception over changes in the provision of environmental services. The participation of government and other institutions (such as landowners represented by *ejidos*[§]) can help to identify critical watersheds for cities, private companies, or non-government organizations. In incipient markets, such as in Mexico, government participation is essential in promoting the type of service most needed for users.

Devising PES programs, such as PSAH, as a rent based on the watershed services preserved (or on the decline in the rate of its loss), necessitates translating ecological functions as measurable and traceable unit of services provided due to the payment (Wunder, 2007). Providing economic incentives to enhance ecosystem service delivery would be ineffective if policies are implemented without tools to differentiate those who alter their management practices in response to the incentive from those free-riders whose behaviors are essentially unaltered (Gilenwater, 2011). To overcome the constraints from the institutional failure, the government must clarify how the service in question and its value will be measured and monitored. We believe that combining market and non-market valuation techniques clarifies the scale of economic distortions due to uncertainty and should help understand the importance of both use and non-use values. The impacts of non-point sources to streamflow can be monitored by establishing a paired-watershed design, which utilizes a calibration period and a control watershed to detect changes in hydrology of a treatment watershed.

[§] Ejidos is one the agrarian reform outcomes generated by the Mexican revolution in the 1920's. As defined by Alcorn and Toledo (1998) an ejido is as an expanse of land, title to which resides in a community of beneficiaries of the Agrarian Reform. Most of the ejidos are collectively owned or cooperatively farmed and the products are also marketed collectively.

Finally, although programs like PSAH are not the panacea to water quality and deforestation problems (Muñoz-Piña et al., 2008), they should be considered in the design of policies for sustainable forest management. PES programs need to reflect the real value of services so providers allocate their maximum effort to internalize the externalities. The real value will come with the use of appropriate economic methods that consider both use and non-use values of watershed services. The involvement of other actors, such as the private sector and non-government organizations, is necessary to improve decision-making and ensure that these kinds of programs achieve their goals.

6. Acknowledgments

We would like to thank CONACYT and IPN for their financial support in the development of this work. CONAFOR provided helpful information regarding the case studies of environmental valuation in Mexico. We are grateful to Daniel Garcia-Hernandez, Celina Perez, and the book Editor, for their inputs in an early version of the manuscript.

7. References

Alcorn, J., Toledo, V. M. (1998). Resilient resource management in Mexico´s forest ecosystems: the contribution of property rights. In F. Berkes and C. Folke, (Eds) *Linking social and ecological systems*. Pages 216–249. Cambridge University Press, Cambridge, UK.

Andreassian, V. (2004). Waters and forests: from historical controversy to scientific debate. *Journal of Hydrology*, 291: 1-27.

Aviles-Polanco, G., Huato-Soberanis, L., Troyo-Dieguez, E., Murillo-Amador, B., Garcia-Hernandez, J.L, Beltran-Morales, L.F. (2010). Valoración Económica del servicio hidrológico del acuífero de La Paz, B.C.S.: Una valoración contingente del uso de agua municipal. *Frontera Norte* 22 (43), 103-128

Boyle, K.J. (2003). Contingent valuation in practice. In Champ, P.A., Boyle, K.J., Brown, T.C. (Eds.) *A primer on nonmarket valuation* (pp. 111-169). Kluger Academic Publishers. Norwell, MA.

Brauman, K.A., Gretchen, C., Duarte, T. K., Mooney, H.M. (2007). The nature and value of ecosystem services: An overview highlighting hydrologic services. *Annual Review of Environmental and Resources*, 32:67–98.

Brown, A. E., L. Zhang, T. A. McMahon, A. W. Western, R. A. Vertessy. (2005). A review of paired catchment studies for determining changes in water yield resulting from alterations in vegetation. *Journal of Hydrology*, 310:28-61.

Carson, R.T., Groves, T. (2007). Incentive and informational properties of preference questions. *Environmental Resource Economics*, 37, 181–210.

Chagoya, J.L., Iglesias, L. (2009). Esquema de pago por servicios ambientales de la Comisión Nacional Forestal, México. In Sepulveda, C., Ibrahim, M. (Editores*). Políticas y sistemas de incentivos para el fomento y adopción de buenas prácticas agrícolas*. Capitulo 10 (pp. 189-204). Centro Agronómico de Investigación y Enseñanza, CATIE. Turrialba, Costa Rica.

Champ, P.A., Boyle, K.J., Brown, T.C. (2003). *A primer on nonmarket valuation*. Kluger Academic Publishers. Norwell, MA.

Chan-Yam, L.B. (2007). *Valoración económica del agua para conocer la disponibilidad de pago en comunidades presentes en ÁPFF Sierra Álamos - Rio Cuchujaqui, Álamos, Sonora*. Tesis. Universidad Autónoma Chapingo. Chapingo, México. 138 p.

Coase, R.H. (1960). The problem of social cost. *The Journal of Law and Economics*, 3(1): 1–44.

CONAFOR (Comisión Nacional Forestal). (2004). Reglas de operación del programa de servicios ambientales hidrológicos. *Diario Oficial de la Federación* (18 de Junio 2004). [Available at http://www.semarnat.gob.mx/leyesynormas/Acuerdos/ACUE_SERV_AMB_HI DRO_18_06_04.pdf, last time accesed June 1, 2011].

Consejo Civil Mexicano. (2008). El Programa de pago por servicios ambientales hidrológicos de la Conafor: revisión crítica y propuestas de modificación. In Pare, L., Robinson, D, Gonzalez, M.A. (Editores). *Gestión de cuencas y servicios ambientales. Perspectivas comunitarias y ciudadanas* (pp. 259-276). INE–SEMARNAT. Mexico, D.F.

Costanza R., d'Arge R., deGroot R.D., Farberk S., Grasso M., Hannon B., Limburg K., Naeem S., O'Neill RV., Paruelo J., Raskin R.G, Suttonkk P., Belt V.D. (1997). The value of the world's ecosystem services and natural capital. *Nature* 387(6630): 253–260.

Daily, G.C. (1997). *Nature's Services: Societal Dependence on Natural Ecosystems*. Island Press: Washington, DC.

DOF (Diario Oficial de la Federación). (2010). Comisión Nacional de los Salarios Mínimos. Vigencia a partir del 1 de Diciembre del 2011. *Diario Oficial de la Federación*. [Available at http://dof.gob.mx/nota_detalle.php?codigo=5172213&fecha=23/12/2010,   Last time accessed: Jun 14, 2011].

Dunkerley, D.L. (2008). Intra-storm evaporation as a component of canopy interception loss in dryland shrub: observations from Fowlers Gap, Australia. *Hydrological Processes*, 22: 1985-1995.

Feen, R.H. (1996). Keeping the balance: ancient Greek philosophical concerns with population and environment. *Population and Environment*, 17(6): 447–458.

Field, B.C. (2008). *Natural resource economics*. Waveland Press, Inc. Long Grove, Illinois.

Freeman III, A.M. (2003). *The measurement of environmental and resource values. Theory and methods*. Resources for the Future Press. Washington, D.C.

Garcia-Angeles, A. (2006). *Valoración económica de los servicios ambientales de Santa Catarina Ixtepeji, Distrito de Ixtlán, Oaxaca*. Tesis. Universidad Autónoma Chapingo. Chapingo, México. 163 p.

Gillenwater. M. 2011.*What is additionality? Part 1: a long standing problem*. Discussion Paper No. 001. GHG Management Institute: Washington, DC. [Available at http://ghginstitute.org/wp-content/uploads/2011/03/AdditionalityPaper_Part-1_ver2_FINAL.pdf, last time accessed Jul 25, 2011].

Gómez-Baggethun E, de Groot R, Lomas PL, Montes C. (2010). The history of ecosystem services in economic theory and practice: from early notions to markets and payment schemes. *Ecological Economics*, 69(6): 1209–1218.

Gutierrez-Villalpando, V. (2006). *Valoración económica del agua potable en la zona urbana de San Cristóbal de las casas, Chiapas*. Tesis de Maestría. El Colegio de la Frontera Sur. San Cristóbal de las Casas, Chis.

Hiscock, K. (2005). *Hydrogeology: principles and practice*. Blackwell Publishing. Malden, MA.

Ikawa, R.; Shimada, J.; Shimizu, T. (2009). Hydrology in mountain regions: observations, processes and dynamics. In Marks, D. Proceedings of the International Commission on Snow and Ice Hydrology: *Hydrology in Mountain Regions: Observations, Processes and Dynamics (pp. 25-33)*. July 2007. Perugia, Italy.

Jackson, C.R., Miwa, M. (2007). Importance of forestry BMPs for water quality. In Proceedings of the *Louisiana Natural Resources Symposium: Human and Other Impacts on Natural Resources–Causes, Quantification, and Implications*, August 13–14, 2007. Louisiana State University, Baton Rouge, pp. 10–31.

Jimenez-Moreno, M.J. (2004). *Valoración de algunos recursos naturales, para conocer la disponibilidad de pago por servicios ambientales en el municipio de Tepetlaoxtoc*. Tesis. Universidad Autónoma Chapingo. Chapingo, México. 75 p.

Locatelli, B., Vignola, R. (2009). Managing watershed services of tropical forests and plantations: Can meta-analyses help? *Forest Ecology and Management*, 258 (9): 1864-1870.

Lopez-Paniagua, C., González-Guillén, M.J, Valdez-Lazalde, J.R, de los Santos H.M. (2007). Demanda, disponibilidad de pago y costo de oportunidad hídrica en la Cuenca Tapalpa, Jalisco. *Madera y Bosques* 13(1): 3-23.

Martin, D.A., Moody, J.A. (2001). Comparison of soil infiltration rates in burned and unburned mountainous watersheds. *Hydrological Processes*, 15: 2893–2903.

McBroom, M. W., R. S. Beasley, M. T. Chang, G. G. Ice. (2008). Storm runoff and sediment losses from forest clearcutting and stand re-establishment with best management practices in east Texas, USA. *Hydrological Processes*, 22:1509-1522.

Monroy-Hernandez, R. (2008). *Valoración económica del servicio ambiental hidrológico en la reserva de la biosfera barranca de Metztitlán, Hidalgo*. Tesis. Universidad Autónoma Chapingo. Chapingo, México. 87 p.

Mitchell, R.C., Carson, R.T. (1989). *Using surveys to value public goods: the contingent valuation method*. Resources for the Future Press. Washington, D.C.

Messerli, B., Viviroli, D., Weingartner, R. (2004). Mountains of the world: vulnerable water towers for the 21st century. *Ambio*, 33(13): 29-34

Muñoz-Piña, C., A. Guevara, J.M. Torres y J. Braña. (2008). Paying for the hydrological services of Mexico's forests: Analysis, negotiations and results. *Ecological Economics*, 65(4):725-736.

Navar, J. (2011). Stemflow variation in Mexico's northeastern forest communities: Its contribution to soil moisture content and aquifer recharge. *Journal of Hydrology*. In press. doi: 10.1016/j.jhydrol.2011.07.006

Neary, D.G., Ice, G.G., Jackson, C.R. (2009). Linkages between forest soils and water quality and quantity. *Forest Ecology and Management*, 258: 2269–2281

Ojeda, M.I., Mayer, A.S., Solomon, B.D. (2008). Economic valuation of environmental services sustained by the Yaqui River Delta. *Ecological Economics*, 65: 155-166.

Orozco-Paredes, L.M. (2006). *Balance hidrológico y valoración económica de la producción de agua en la microcuenca del Río Zahuapan, Tlaxco, Tlax*. Tesis. Universidad Autónoma Chapingo. Chapingo, México. 174 p.

Pagiola, E., Bishop, J., Landel-Mills, N. (2003). *La venta de servicios ambientales forestales. Mecanismos basados en el mercado para la conservación y el desarrollo.* INE-SEMARNAT. Mexico, D.F.

Paré, L, Robinson, D., Gonzalez, M.A. (2008). *Gestión de cuencas y servicios ambientales perspectivas comunitarias y ciudadanas.* INE-SEMARNAT. Mexico, D.F.

Pattanayak, S.K. (2004). Valuing watershed services: concepts and empirics from southeast Asia. *Agriculture, Ecosystems and Environment,* 104: 171–184

Pearce, D.W. (2001). The economic value of forest ecosystems. *Ecosystem Health,* 7(4): 284–296.

Perez-Verdin, G., Kim, Y-S., Hospodarsky, D., Tecle, A. (2009). Factors driving deforestation in common-pool resources in northern Mexico. *Journal of Environmental Management,* 90:331-340

Pierson, F.B., Robichaud, P.R., Moffet, C.A., Spaeth, K.E., Williams, C.J., Hardegree, S.P., Clark, P.E. (2008). Soil water repellency and infiltration in coarse-textured soils of burned and unburned sagebrush ecosystems. *Catena,* 74: 98–108.

Plottu, E, Plottu, B. (2007). The concept of total economic value of environment: a reconsideration within a hierarchical rationality. *Ecological Economics,* 61: 52-61.

Sanjurjo, E. (2006). *Aplicación de la metodología de valoración contingente para determinar el valor que asignan los habitantes de San Luís Río Colorado a la existencia de flujos de agua en la zona del Delta del Río Colorado.* Dirección de Economía Ambiental. INE-SEMARNAT. Mexico, D.F.

Schlapfer, F. (2008). Contingent valuation: a new perspective. *Ecological Economics,* 64: 729–740

Silva-Flores, R., Perez-Verdin, G., Navar-Chaidez, J.J. (2010). Valoración económica de los servicios ambientales hidrológicos en El Salto, P.N., Durango. *Madera y Bosques,* 16(1): 31-49.

Soto, M.G, Bateman, I.J. (2006). Scope sensitivity in households' willingness to pay for maintained and improved water supplies in a developing world urban area: Investigating the influence of baseline supply quality and income distribution upon stated preferences in Mexico City. *Water Resources Research,* 42: 1-15.

Swank, W.T., Swift, L.W., and Douglass, J.E. (1988). Streamflow changes associated with forest cutting species conversions, and natural disturbances. In *Forest Hydrology and Ecology at Coweeta.* W.T. Swank and D.A. Crossley (Editors). pp 297-312. Springer Verlag, New York.

TEEB (The Economics of Ecosystem and Biodiversity). (2010). The economics of valuing ecosystem services and biodiversity. In *The Economics of Ecosystems and Biodiversity: Ecological and Economic Foundations,* Kumar P (ed). (Chapter 5, 133 p). UNEP, Bonn, Germany.

Turner, R.K, Daily, G.C. (2008). The ecosystem services framework and natural capital conservation. *Environmental and Resource Economics,* 39(1): 25–35.

Vasquez, W.F., Mozumder, P., Hernandez-Arce, J, Berrens, R.P. (2009). Willingness to pay for safe drinking water: evidence from Parral, Mexico. *Journal of Environmental Management,* 90: 3391–3400.

Ward, A.D., Trimble, S.W. (2004). *Environmental hydrology.* CRC Press. Boca Raton, Florida.

Whittington, D. (2002). Improving the performance of contingent valuation studies in developing countries. *Environmental and Resource Economics*, 22: 323–367, 2002.

Willmott, C.J, Feddema, J.J. (1992). A more rational climatic moisture index. *The Professional Geographer*, 44(1): 84 – 88.

Wunder, S. (2007). The efficiency of payments for environmental services in tropical conservation. *Conservation Biology*, 21(1): 48–58.

Market-Based Approaches Toward the Development of Urban Forest Carbon Projects in the United States

Neelam C. Poudyal[1], Jacek P. Siry[1] and J. M. Bowker[2]
[1]Warnell School of Forestry & Natural Resources, University of Georgia, Athens, GA
[2]USDA Forest Service, Southern Research Station, Athens, GA
USA

1. Introduction

1.1 Urban forestry in the United States: Status and scope

The United States has observed unprecedented urban growth over the last few decades. Nowak et al. (2005) noted that between 1990 and 2000, the share of urban land area in the nation increased from 2.5% to 3.1%. Existing urban areas in the U.S. maintain average tree coverage of 27% (Nowak et al. 2001), and consist of millions of trees along streets and in parks, riparian buffers, and other public areas. Further, Walton and Nowak (2005) predicted that this urban area will continue to expand through 2050, eventually covering up to 8.1% of the country's area. Some of the expected urban development will come at the expense of currently forested areas. This may further the scope of afforestation and subsequent reforestation as part of urban forest management.

Increasing with the area of urban land is the geographical coverage of urban forests. Urban areas nationwide support more than 3.8 billion trees (Nowak et al. 2002), whereas as many as 70 billion trees are estimated to be growing in the urban and urbanizing areas throughout the nation (Bratkovich et al. 2008). A brief look at urban tree inventory data at individual state and city levels confirms that urban trees are a significant component of forest resources at local and regional levels. Table 1 presents canopy coverage and tree inventory data for five selected states and cities to illustrate the relative stocking of urban trees at individual state and municipal level (Nowak et al. 2001). Some of the states have smaller urban canopy coverage, but are densely stocked. Recent urban forest inventories also suggest that there is substantial variation of tree stocks among the United States cities, which ranges from roughly 15 trees per acre in Jersey City, New Jersey to about 113 trees per acre in Atlanta, Georgia (Nowak et al. 2010).

1.2 Issues facing urban forestry in the United States

Sustainable management of forest resources nationwide, regardless of their ownership and management objectives, is facing a number of challenges. Urban forestry is no exception. Sustainable forest management implies conservation and sustainable use of forest resources across all ownerships including urban forests, which are typically managed by local governments (e.g., municipality, city, metropolitan council, town). While population growth

State	Urban tree cover (%)	Urban trees (thousands)	City	Urban tree cover (%)	Urban trees (thousands)
Georgia	55.3	232,906	Atlanta	36.7	9,420
Alabama	48.2	205,847	Boston	22.3	1,180
Ohio	38.3	191,113	Baltimore	21.5	2,600
Florida	18.4	169,587	Oakland	21.0	1,590
Tennessee	43.9	163,783	New York	20.9	5,220

Note: Adopted from Nowak et al. (2010,p. 39)

Table 1. Tree cover and number of trees for selected U.S. states and cities

and development pressures accelerate the loss of wild lands and expansion of urban and suburban areas, protecting and managing trees for a variety of societal and environmental benefits often remains up to local governments. Urban forest management in the United States and elsewhere is facing substantial challenges which threaten the long-term conservation and management of urban tree and park resources. Major factors currently under consideration in the U.S. include the following:

• Disease and pest infestation
• Invasive species
• Wildfires
• Heavy recreational use
• Fragmentation
• Air pollution
• Lack of community participation
• Insufficient funding

Recent forest disturbance research (Holmes et al., 2008) illustrates a range of biological and socio-economic threats to the United States forest systems. A number of invasive stem borers and sap sucking pests such as Emerald Ash Borer, Gypsy Moth, Hemlock Woolly Adelgid have already killed thousands of trees of high amenity and ecological value. A number of exotic plant species including Kudzu, Chinese Privet, and English Ivy have invaded native landscapes in urban parks and roadside plantations. Increasing air pollution due to auto emissions and atmospheric pollution from industrial plants that are often located near urban areas have negatively affected the physiology and ecology of urban landscapes. Furthermore, with rapid population growth, per capita public open space is declining and existing urban forest resources in some areas are being ecologically destroyed due to heavy use (Poudyal et al. 2009). On the other hand, garnering sufficient community participation in urban tree management is challenging due to changing socio-demographics and ethnic heterogeneity in major metropolitan areas. Residents living in a heterogeneous community usually show varying levels of interest towards the maintenance and management of community resources like urban trees, which makes planning and implementation complicated (Gaither et al. 2011).

Another big challenge facing urban forestry right now is insufficient funding. A perennial source of income could greatly contribute to making the urban forest programs financially self-sufficient and sustainable. Indeed, with sufficient funding, local governments could put together efforts aimed at managing many of the other issues listed above. This is why it is important to address the marketability and revenue generating potential of ecosystem services that urban forests provide.

1.3 Towards a financially self-reliant urban forestry

As stated in the preceding section, local government budget problems, and the lack of adequate funding for tree care and maintenance has been considered a major issue in the United States. Mere tree planting along roadsides or on vacant lots within city limits does not define urban forestry. Rather, it involves tree care and maintenance and management (e.g., pruning, clearing, disposal), for which about two-thirds of an urban forest project budget needs to be typically allocated (American Public Works Association, 2007). However, urban forest projects during tough economic times are often overlooked when setting funding and management priorities. Private individuals, albeit usually appreciative of the amenity benefits of urban trees, do not always support the 'tax approach' to finance tree care and management programs. Urban forests bear some characteristics of 'public goods,' meaning that once an output or service is supplied, nobody can be effectively excluded from enjoying it, thereby leading to free-rider problems (Freeman, 2003). Private firms and for-profit organizations have few incentives to provide and maintain such resource. Therefore, if the good or service is to be provided, government must play a major role, either by direct provision or by providing incentives to the private sector.

The sustainable management of urban trees will require continuous funding and a reliable and well-established income generating mechanism at local level. The Urban and Community Forestry Program of United States Department of Agriculture aims at enabling the development of self-sufficient local urban and community forestry programs nationwide. As the provision of a range of public services and basic infrastructure compete for tax revenue, local governments are required to look for external sources of funding to keep their urban forestry programs operating adequately. In many cases, forest management programs, regardless of their location and ownership, will not be sustainable unless they are financially self-sufficient.

Because of the aesthetic and amenity purposes of urban forest management, neither timber neither timber harvesting nor planting of fast-growing cash tree crops are compatible options, or even a debatable alternatives. However, among a wide range of ecosystem services, carbon sequestration is especially promising. Nowak & Crane (2002) estimated that urban forests in the conterminous United States can store 770 million tons of atmospheric carbon, valued at $14.3 billion, assuming conversion to tradable carbon credits and then-current prices. Translating those numbers into annual terms, the United States urban forests absorb nearly 23 million tons of carbon, which can generate $460 million in revenue -- again assuming conversion to tradable carbon credits and concurrent prices. By appropriately managing urban trees and forests for maximum carbon sequestration, cities can collect revenue from selling credits for carbon absorbed and stored in urban trees. Revenue generated in this manner will not strain local tax revenue collections, and will help fund sustainable urban forest management. Given the fact that markets for carbon offset credits have recently emerged, carbon credits become worth investigating.

Federal and state agencies are trying to promote carbon trading in community and urban forestry as evidenced by a series of recently published policy documents. For example, a recently released USDA Forest Service document on open space conservation strategy has listed promotion of market-based approaches to enhance carbon-credit trading as one of the top thirteen priority actions (USDA Forest Service, 2007). Despite its significant potential and increasing policy emphasis, the market for urban forest carbon credits has not been well developed. This outcome in part is a result of the lack of appropriate and broadly accepted market protocols, and the limited understanding of entrepreneurial principles associated with this product. Developing carbon markets will require a thorough understanding of the preferences and expectations of potential buyers per the characteristics, quality, and price of carbon credits. It will also require information about the technical and managerial capacities of the potential sellers to develop carbon offset projects. This chapter highlights some of the findings of a recently completed comprehensive research project in the United States that examined the capacities, interests, and expectations of both the potential sellers and buyers of carbon credits generated from urban forest projects.

2. Objective

The objective of the material presented in this chapter is to address the feasibility of establishing a market for urban forest carbon credits. This will be achieved by assessing the interest of key stakeholders involved in potential market for this output. Stakeholders' perspectives will be discussed in a broader context of making urban forestry a source of carbon credits that will help make it financially self-sufficient and sustainable.

3. Approach

The project started with the identification of key stakeholders in a potential market for urban forest carbon credits. In order to establish a market, potential buyers and sellers of the urban carbon credits must be identified. Given the nature of ownership, local governments and municipalities were considered as the sellers of urban forest credits. A web-based survey was implemented during 2007-2008, contacting urban foresters, arborists and other officials responsible for overseeing their urban forest. Contact details of those officials were obtained from the Society of Municipal Arborists (SMA). The survey questionnaire focused on cities' current urban forest information and management practices, existing stock and available technical and managerial expertise, and interest in participating in an urban forest carbon offset trading program.

Identifying the potential buyers was challenging given that the United States market for forest urban carbon credits has not been well developed. However, because credit buyers in the United States are voluntarily participating in carbon trading rather than complying with mandatory government regulations, existing credit buyers may have unique preferences for credits sourced from specific locations such as urban forests. Therefore, businesses and organizations that are currently participating in carbon markets were identified as the potential buyers of urban forest credits. While many buyers purchase carbon credits from over-the-counter (OTC) market, surveying them is difficult due to the lack of their contact information. For this reason, primary buyers of carbon credits at the Chicago Climate Exchange (CCX), which was the largest carbon trading platform in North America, were surveyed as the potential buyers.

All CCX members and associate members were invited to complete a survey that covered questions regarding their attitudes and perceptions related to climate change, government regulation of greenhouse gas emissions, and their preferences for credits sourced from a variety of carbon project types, including urban forestry. Some of the questions were related to their willingness to purchase urban forest carbon credits and the price they were willing to pay. This survey was conducted during late 2009.

4. Key observations

This section presents some basic statistics and summary of survey responses from the surveys of both the buyers and sellers.

4.1 Seller's survey

From a total of 277 successfully delivered surveys, an adjusted response rate of 54% was achieved. The group of responding municipalities was highly diverse in terms of population size and regional location. About one-fifth of respondents in the sample represented large cities (with population larger than 100,000) and another one-fifth represented small cities (population less than 20,000). Roughly one-third of the respondents were from mid-size cities (with population between 20,000 and 50,000). Respondents from the Northeast region were slightly underrepresented (6%) while other regions (i.e., Midwest, 37%; South, 27%; and West, 31%) were more uniformly represented. Only one-fifth of the respondents were familiar with the Chicago Climate Exchange which, at the time of survey, was the only actively operating carbon trading platform in the country. Further details on respondent's characteristics can be found in Poudyal et al., (2010).

Local government units that responded to the survey indicated that they were maintaining or managing urban forest resources of some sort within their jurisdiction. The exact form of urban forests varied from urban parks, forest patches within city limits to individual trees along streets, roadside tree plantings and protected vegetation along critical riparian buffer areas. More importantly, a clear majority of responding municipalities (63%) had an official designated to oversee the urban tree care and management activities. Similarly, about 56% of the respondents had at least a portion of their forest resource recently inventoried. A similar survey of U.S. cities recently conducted by the United States Conference of Mayors suggested that as much as 55% of cities had a current inventory of urban tree canopy (Nowak et al., 2010).

When asked if local governments were currently participating in any climate change initiatives, respondents identified a number of projects, including remodeling and construction of energy efficient buildings, using alternative fuel vehicles, capturing landfill methane, and planting trees. More importantly, tree planting was the most common initiative undertaken recently (85% of the respondents) to help mitigate climate change (Figure 1). Similarly, about 50% in the sample indicated either using alternative fuel vehicles or constructing/remodeling energy efficient buildings as a recently undertaken initiative to mitigate climate change. It seems that local governments' tree plantation investments in recent years, and perhaps in the near future, would give them an advantage in initiating active-management-based urban forest offset projects. This is necessitated because the already planted stocks do not meet the 'additionality' criterion, unless they are placed under an intensive management regime to boost their carbon sequestration rate.

Prior to reading the questionnaire, approximately one-third of the respondents were familiar with the idea of carbon storage and offset selling. However, very few of the responding municipalities were familiar with existing market platforms like the Chicago Climate Exchange where they could sell their carbon credits. When asked if their city would be willing to participate in a carbon offset selling scheme, 29 out of 150 (roughly 20%) indicated that they were interested or very interested in such a program. On the other hand, 15 respondents (about 2%) indicated that their city was uninterested or not at all interested in carbon trading at this point. An econometric model was estimated to examine factors that

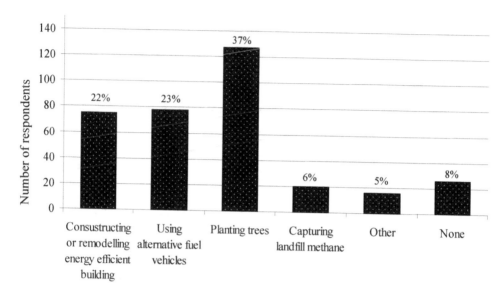

Fig. 1. Number of municipal governments currently participating in various climate change mitigation initiatives

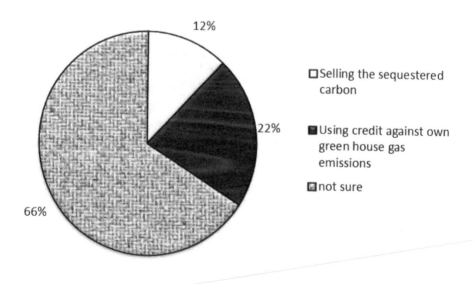

Fig. 2. Local government's plan to utilize the certified carbon credits sourced from their urban forests

influenced respondent's willingness to participate in carbon trading program. Detailed results in Poudyal et al., (2010) indicate that a local government's decision to participate in carbon trading was positively influenced by staff's knowledge of carbon sequestration and familiarity with carbon trading intuitions such as CCX, potential interest of voters, level of urbanization, and a city's need for generating revenue. This observation indicates that along with the increasing need of local governments to generate revenue combined with rising environmental awareness of voters and urban congestion, more local government units will be interested in selling carbon credits though urban forest projects.

Cities which were yet to generate certified offset credits were asked about their plans for using their credits. A majority (66%) were unsure, which is a common response for such a hypothetical question (Figure 2). Among the remaining one-third who had tentative plans regarding the utilization of their certified credits, a significantly higher number of respondents (22%) indicated that they will count the credits against the city government's green house gas emissions rather than selling them to interested buyers (12%). Hence, as the public pressure grows for environmental compliance, and as government units require more credits to offset their own emissions, some local governments may have fewer credits left to sell in the market. How these currently 'unsure' respondents will decide the use of their carbon credits could largely determine whether this may become an issue at all.

4.2 Buyer's survey

From a total of 155 successfully delivered addresses, an adjusted response rate of 41% was achieved. Respondent businesses and organizations (i.e., members and associate members at the CCX) were diverse in terms of their business characteristics such as profit motive, employment and geographical scope of business operations. Slightly more than half in the sample were private or for-profit organizations, whereas just about a quarter of the sample were public or non-governmental organizations. The remaining one-fifth were government institutions. About half of them confined their business operations to the United States. About one-half of all respondents had a target of reducing their greenhouse gas emissions by 5% in the near future. In terms of their carbon trading history, one-half of the sample had been participating in carbon trading for 3 or more years. Respondents, on average, purchased about thirty three thousand metric ton equivalents of carbon dioxide offset credits in the most recent calendar year (i.e., 2008). Further details of respondents' characteristics can be found in Poudyal et al., (2011).

Overall, current buyers of carbon credits in the North American market were found to be pro-environmental and generally supportive of government regulation to control the greenhouse gas emissions. Discussing buyer attributes in detail is beyond the scope of this chapter, but a rigorous analysis of their responses can be found in Poudyal et al. (2011). Buyers were asked to rank credit types by the location of an offset project. Respondents showed much higher preference for credits sourced from local projects than those generated from regional or international projects (Figure 3). Since a number of businesses and organizations interested in offsetting their emissions are located around urban areas, a noticeably higher preference for locally generated credits shows a potentially high value of such credits to buyers.

A more specific question required respondents to rank carbon credits generated from different sources. As Figure 4 shows, buyers clearly placed the highest value on the credits sourced from renewable energy projects. However, their preference for urban forest credits was relatively higher than those sourced from agriculture or methane soil projects. Urban forest credits were found as desirable as rural forestry credits among the credit buyers in the North American market.

Figure 4 suggests that urban forest carbon credits may be fairly competitive in the market. However, whether they will generate more revenue compared to other credit types is a separate question. Buyers' responses in terms of willingness to offer a premium for specific credit types varied substantially among various types of projects. In addition to urban forest credits, respondents were asked to consider offering premiums for credits sourced from three other types of projects: (1) projects promoting nature conservation in developing countries; (2) projects aimed at alleviating poverty in developing countries through carbon payment to forest landowners; and (3) rural forest projects in the United States. While a modest (roughly 15%) number of respondents consistently rejected the idea of paying premium for any kind of carbon credits, many respondents had favored offering a premium for credits sourced from a range of projects. Among the projects listed above, roughly 55% of respondents indicated that they would be willing to pay a premium for urban forest credits. None of the other projects generated a higher level of support or willingness to offer premium. Compared to the current market price of credits for which the source is not generally disclosed, urban forest credits, if known, could draw a significant premium.

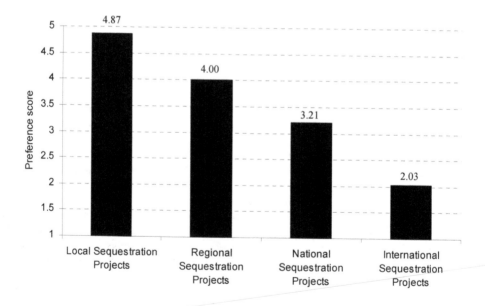

Fig. 3. Buyers' preference for carbon credits by project location

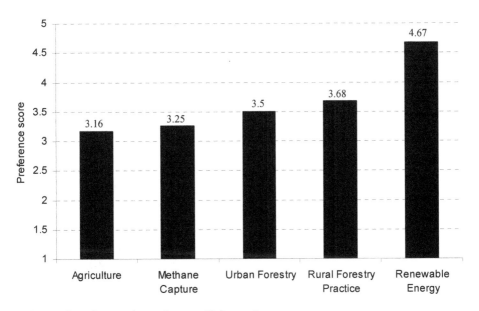

Fig. 4. Buyers' preference for carbon credit by project types

5. Concluding remarks

Buyers and sellers of carbon offsets are interested in this new urban forest output. Urban forest credits are more desirable than other types of credits and buyers are willing to pay a higher price. This will certainly help local governments to be more competitive in the offset market. In fact, this could present an opportunity to be active in localized markets and generate sufficient revenues while preserving urban forests in the long-run and providing a wide range of co-benefits to the society. We argue that promising financial potential provides incentives for local governments to utilize marginal and abandoned industrial lands to increase urban canopy coverage, and to adopt stricter tree management ordinances to boost the carbon storage capacity of public trees. Nowak et al. (2010) noted that about one half of the sample in a recent survey of the United States cities with population of 30,000 or more indicated that expanding tree canopy is their goal and as much as 95% of them have even adopted some sort of tree management ordinance (City Policy Associates, 2008). Current local government initiatives are not necessarily motivated by the need to develop an offset market, but these recent developments when considered together with our results suggest that local governments adopting such policy initiatives may have an advantage with early entrance into the carbon market. Thanks to a number of federal programs that currently offer federal funds to help local communities establish sustainable, clean and green communities, local governments could establish such innovative projects. The Climate Showcase Community Grants of the US Environmental Protection Agency, Sustainable Communities Grants of US Department of Housing and Urban Development, and US Department of Energy's Energy Efficiently and Conservation Block Grants are just a few examples (American Public Works Association, 2007).

However, some research results suggest that the long-term viability of urban forests as a source of carbon credit may be debatable. First, as Nowak et al., (2010) note that increasing tree coverage may increase the potential for storing additional carbon in urban tress, but the maximum tree coverage will entail additional risk and costs, such as wildlife risk along high density residential areas, human-wildlife conflict due to expanded habitat for birds and animal species, and water usage. A long-term strategy for optimizing the social, economic and ecological benefits might be needed to make this effort sustainable. Second, researchers are still debating the net carbon footprint of urban forest projects themselves. Third, our results suggest that more municipalities are likely to use their offset credits against their own emissions targets if they have to comply with a mandatory emission reduction regulations in the future. As more cities sign the Mayors Climate Change Protection Agreement, larger number of carbon offsets will have to be used by cities themselves to improve their green image and meet their constituents' environmental expectations. But again, whether this issue will remain a real concern will largely depend on how the interest and responses of the currently "unsure" group will unfold against increasing demand for carbon credit in future.

Nevertheless, given some of the unique characteristics of urban forest, cities could still produce surplus and market offset credits. Nowak and Crane (2002) argued that by fostering larger trees and by inducing energy savings effects, an urban tree may store four times more carbon than a single tree in a forest stand. However, this assertion should be viewed cautiously as it was derived from a simulation study rather than an empirical measurement of actual sequestration between urban trees and its rural counterparts.

In any case, it seems that there are increasing signs of favorable views and interest among administrators and urban forestry professionals to initiate projects generating carbon offsets. For example, our observations of sellers' motivations and their interests corroborates the findings from a recent survey of members of Society of Municipal Arborists, in which researchers observed that urban forestry professionals are embracing ecosystem services such as climate management, habitat protection, and biodiversity conservation as departmental goals beyond their traditional focus on enhancing property values and protecting utility lines (Young, 2010). It is reasonable to assume that there might be a shift in the way both residents and city managers view the significance and utility of urban forest resources. Part of the enthusiasm and favorable view of professionals probably relies on the availability of practical and user-friendly computer models such as i-Tree or UFore (http://www.itreetools.org) that are useful in quantifying and valuing city forests' offset capacity. All these factors broaden the scope of future urban forest management to include benefits like carbon offset credits.

Key findings highlighted in this chapter provide a holistic view of the market potential and opportunities for making urban forest projects financially self-reliant and more sustainable. Of specific interest to stakeholders are the deeper understanding of the preferences, motivations, and expectations of potential players in the context of establishing markets for urban forest carbon credits. This information could be used to develop new and expand existing market protocols for carbon credits sourced from urban forestry projects.

While this study was based in the United States, the challenge of generating income from urban and community forest projects is likely transferable to other developed countries. Accordingly, many local governments outside the United States are also working to measure and quantify carbon credits generated by their urban forests. While European and Scandinavian countries are already leading in several climate and carbon offset initiatives, some Asian countries (Liu & Li, 2011) and African countries (Stoffberg et al., 2010) have also begun quantification and valuation of carbon sequestration in their urban forests. As more cities and local governments look for ways to make their urban forest projects financially self-sufficient and sustainable, policy implications and recommendations available in this chapter and associated publications should be useful in guiding urban forest management in the United States and beyond.

6. References

American Public Works Association. (2007). Urban forestry best management practices for public works managers: Budgeting and funding. American Public Works Association Press, 20p.

City Policy Associates. (2008). Protecting and developing the urban tree canopy: a 135-city survey. United States Conference of Mayors. Washington, DC.34p.

Freeman, M., Herriges, J. A., and C. L. King. (2003). *The measurement of environmental and resource values: Theory and methods*, Second Edition. Resources for the Future Press, 420p, ISBN 978-0915707690

Holmes, T. P., Prestemon, J. P., & Abt, K. L. (2008). *The economics of forest disturbances: wildfire, storms and invasive species*. Springer Publisher, 422p, ISSN 978-90-481-7115-6

Gaither, C. J., Poudyal, N. C., Goodrick, S., Bowker, J. M., Malone, S., & Gan, J.(2011). Wildland fire risk and social vulnerability in the Southeastern United States: An exploratory spatial data analysis approach. *Forest Policy and Economics*, Vol.13, 24-36, ISSN 1389-9341.

Liu, C., & Li, X. (2011). Carbon sequestration by urban forests in Shenyang, China. *Urban Forestry and Urban Greening*. In Press, ISSN 1618-8667.

Nowak, D. J., Walton, J. T., Dwyer, J. F. , Kaya, L. G. , & Myeong, S. (2005). The increasing influence of urban environments on US forest management. *Journal of Forestry*, Vol. 103 (8), 377-382, ISSN 0022-1201.

Nowak, D. J., & Crane, D. E. (2002). Carbon storage and sequestration by urban trees in the USA. *Environmental Pollution*, Vol. 116,381-389, ISSN 0269-7491.

Nowak, D. J., Noble, M. H., Sisinni, S. M., & Dwyer, J. F. (2001). Assessing the US urban forest resource. *Journal of Forestry* Vol. 99 (3), 37-42, ISSN 0022-1201.

Nowak, D. J., Novle, M. H., Sisinni, S. M. , & Dwyer, F. J. (2001). People and trees: Assessing the US urban forest resources. *Journal of Forestry*, Vol. 99(3):37-42, ISSN 0022-1201.

Nowak, D. J., Stein, S. M., Randler, P. B., Greenfield, E. J., Comas, S. J., Carr, M. A., & Alig, R. J. (2010). *Sustaining America's urban trees and forests*. General Technical Report NRS-62. Newton Square, PA: USDA Forest Service, Northern Research Station, 27p.

Poudyal, N. C., Hodges, D. G., & Merrett, C. D.(2009). A hedonic analysis of demand for and benefit of urban recreation parks. *Land Use Policy*, Vol. 26(4), 975-983, ISSN 0264-8377.

Poudyal, N. C., Siry, J. P., & Bowker, J. M. (2011). Urban forests and carbon credits: Credit buyer's perspectives. *Journal of Forestry*, In Press, ISSN 0022-1201.

Poudyal, N. C., Siry, J. P., & Bowker, J. M. (2010). Urban forestry's potential to supply marketable carbon credits: A survey of local governments in the United States. *Forest Policy and Economics*, Vol. 12, 432-438, ISSN 1389-9341.

Stoffberg, G. H., van Rooyen, M .W., van der Linde, M. J., & Groenveld, H. T. (2010). Carbon sequestration estimates of indigenous street trees in the City of Tshwane, South Africa. *Urban Forestry and Urban Greening,* Vol. 9 (1): 9-14, ISSN 1618-8667.

USDA Forest Service.(2007). *Forest Service open space conservation strategy: cooperating across boundaries to sustain working and natural landscapes.* General Technical Report, United States Department of Agriculture, Forest Service, FS-889, 16p.

Young, R. F. (2010). Managing municipal green space for ecosystem services. *Urban Forestry and Urban Greening,* Vol. 9 (4), 313-321, ISSN 1618-8667.

3

Multiple Services from Alpine Forests and Policies for Local Development*

Ilaria Goio[1], Geremia Gios[1], Rocco Scolozzi[2] and Alessandro Gretter[2]
[1]Department of Economics, University of Trento
[2]Research and Innovation Centre, Fondazione Edmund Mach –Michele all'Adige (Trento)
Italy

1. Introduction

The starting point of the analysis here presented is the concept of ecosystem services, which could help us appreciate natural systems as vital assets, recognizing the central roles that they play in supporting human well-being, either at the local or global level. In fact, ecosystem services provide benefits, in terms of goods and services, both to people living in the mountains and to people living outside them. At the moment, these services are seriously threatened, and "their global degradation is increasingly jeopardizing development goals"(OECD, 2008). As a consequence, it is necessary to reverse this trend while, at the same time, meeting the increasing demands of and interests in such services.[1]

The focus of our study are the alpine forest ecosystems, which represent a fundamental resource for people living in mountain areas and for human society, in general.[2] In fact, it is commonly known that forests nowadays fulfil several other functions, in addition to what has been perceived as their main function (the productive one). These functions include the protective function, the landscape and recreational function and the ecological function. This functionality means that forests not only produce goods but also various social and environmental services,[3] contributing, in many different ways, to the welfare of humans. This capacity is well summarized in the concept of "multi-functionality". It is clear that "better understanding of the full range of goods and services supplied by forests is essential for optimal utilization of forests, and it may provide an economic rationale for sustainable forestry" (Lange, 2004).

---

* This paper is the result of its authors' common reflections. However, single sections have been written, as follows: Ilaria Goio wrote 1, 3, 4.1 and 6.1, Geremia Gios wrote 4, 5 and 6; Rocco Scolozzi wrote 2; and Alessandro Gretter wrote 6.2.

[1] "One of the most important problems that our society currently faces is how to strike a suitable balance between the conversion of natural capital to economic production and its conservation to provide ecosystem services" (Farley & Costanza, 2010).

[2] According to the Millennium Ecosystem Assessment (MEA, 2005) the "environmental conservation and sustainable land use in the world's mountains are not only a necessary condition for sustainable local livelihoods, but also for well-being of nearly half the world's population who live downstream and depend on mountain resources".

[3] Historically, the nature and value of these services have largely been ignored until their disruption or loss has highlighted their importance (Daily et al. 1997).

Within this framework, the main objective of this work is to define the management policies that allow efficient and effective use of goods and services produced by forests.

Clearly, these policies will differ in relation to the kinds of goods or services considered and also in relation to the specific socio-economic and environmental context of a given area. In particular, in our analysis, we will make reference to the landscape and recreational function and to its economic assessment, as partly learned by our working experience in the Alpine context of the Autonomous Province of Trento (Italy).

We would like to suggest to the public and local policy makers of the southern part of the Alps, some general economic policy instruments. The objective of these policy instruments is twofold: on the one hand, they permit policy makers the use ("with a sufficient flexibility in order to operate within constantly changing circumstances" [OECD, 1999]) of the above-mentioned goods and services. On the other hand, these policy instruments provide a useful support for orienting their action towards a territorial policy, which is able to give, from the perspective of sustainable development,[4] equal justice to the economic, social, and environmental components of forests.

Within the process just outlined, a key role is played by the local and non-local stakeholders. That is, some stakeholders are the actors who provide environmental benefits, and therefore, have to be remunerated. Other stakeholders should pay for taking advantage of the environmental benefits. For this reason it is necessary to understand how the cited actors perceive the factors connected with sustainability, facilitating and promoting an enriching exchange of views, knowledge and initiatives. To provide a complete and reliable overview, these point of view exchanges should involve both public and private actors, creating new synergies and new partnerships in the area.

This chapter is structured as follows. The next section explains the principal characteristics of ecosystem services and section 3 provides some considerations about the multi-functionality of forests. In section 4, specific reflections on forest joint-productions are presented, including brief considerations of market factors and payment for ecosystem services. Section 5 focuses on the particular case of landscape and recreational services and section 6 illustrates some policy implications for public decision makers in the alpine areas, with particular reference to the need for a participative approach, and offers evidence from Alpine examples.

2. Ecosystem services from mountain areas

Ecosystems are complex systems that provide humanity with vital services through interacting ecological processes. With regard to mountain areas, forest ecosystems provide wood products and a wide range of non-wood products and services, e.g., regulation of the climate and water supply, purification of the air and drinking water, protection against soil

---

[4] The Brundtland Commission's report, published in 1987, defined sustainable development as "development which meets the needs of current generations without compromising the ability of future generations to meet their own needs" (Brundtland, 1987). Recently, the Research Institute for Humanity and Nature (Kyoto, Japan), proposed a reinterpretation of the sustainable development concept, referring to the idea of futurability. "Sustainability is a static and conservative concept that focuses on the continuation of the present-day anthroposphere (i.e., *sustainable parasitism*), although dynamic co-evolution between human and nature could be an alternative definition" (Newman, 2005). "In contrast, futurability is a more dynamic and ambitious concept that seeks truly sustainable and futurable human–environment interactions, namely *futurable mutualism*" (Handoh & Hidaka, 2010).

erosion and the support of soil fertility. Forest ecosystems also play an important role in the aesthetic and recreational values of landscapes, supporting increasing worldwide tourism. Studies conducted since the 1980s indicate that forest values may be much higher than timber values per hectare (Peters et al., 1989). As a consequence, there has been an increasing realization that many other products and services generated by forests are essential to the well being of local communities and are required by society at large. In particular, the FAO (2005) defined non-wood forest products as products and goods "that are tangible, of biological origin other than wood, derived from forests, other wooded land and trees outside forests." These non-wood forest products include mushrooms, fruit, leaves, plants and animals collected or grown in forests, and they are used as food, fodder, medicine and raw materials for handicrafts. They have significance as cultural objects and as a source of income. This definition of non-wood forest products neglects intangible forest services (e.g., ecotourism, bio prospecting) and forest benefits (e.g., soil conservation, watershed protection and maintenance of biodiversity), which are clearly more difficult to assess and quantify than goods. Therefore, a new open system of terms for forest-dependent resources was proposed (Mantau et al., 2001, 2007): "Forest Goods and Services (FOGS), defined as resources of biological origin, associated with forests, other wooded land and trees outside forests".

Specific typologies were proposed to describe the forest transactions (uses) of interest to facilitate analyses or marketing. They consider three basic levels: 1. resource, 2. product and 3. user.[5]

Each of these levels may be internally classified into many hierarchical levels. For example, the "resource" plant may provide a "product" such as erosion control for the "user" state, but may also offer a different "product," such as fuel wood to the "user" local community. In effect, each resource may be structured into several products, and these products, in turn, are handled and consumed by many different user groups. A systematic taxonomy definition of goods and services (Mantau et al., 2001) may help in the examination and description of the value chains that are increasingly being developed as a basis for interventions to promote successful commercialization of FOGS.

Besides the FOGS, as defined above, the concept of ecosystem services better recognizes potential values for ecological/ecosystem processes *per se*. The MEA (2005) breaks ecosystem services into four different classes:

- *Provisioning services,* which are the products obtained from ecosystems, including food, fiber, fuel, genetic resources, ornamental resources, freshwater, biochemical, natural medicines, and pharmaceuticals.

---

[5] In more depth:

- Resource: in the context of the forest, anything of biological origin that is of use to humans and the basis for any output. For instance, resources for goods are energy, carbon, land, water, materials, plants, foodstuff, fibre, medicine, extractives and live plants or animals.
- Product: anything that can be offered to a market that might satisfy a want or need. A product can be a simple marketable good (e.g., fuel wood) or service (e.g., recreation) or combination of both (i.e., composite products or commodities, such as Christmas tree markets and guided mushroom-picking walks).
- User: any group of people that benefits from a product. This category includes collectors, processors, middlemen, retailers and the end-user or client. It therefore describes the market or value-chain for a given product.

- *Regulating services*, which are the benefits obtained from the regulation of ecosystem processes, including air quality regulation, climate regulation, water regulation, erosion regulation, water purification and waste treatment, disease regulation, pest regulation, pollination, and natural hazard regulation.
- *Cultural services*, which are the non-material benefits people obtain from ecosystems through spiritual enrichment, cognitive development, reflection, recreation, and aesthetic experiences, including cultural diversity, spiritual and religious values, knowledge systems, educational values, inspiration, aesthetic values, social relations, sense of place, cultural heritage values, recreation, and ecotourism.
- *Supporting services*, which are necessary for the production of all other ecosystem services.

Figure 1 summarizes well the different classes with reference to the mountain ecosystems. Mountains and their ecosystems provide all services from each of the four main MEA categories, as widely documented in the RUBICODE project.[6]

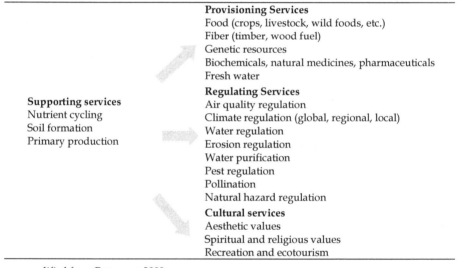

Supporting services
Nutrient cycling
Soil formation
Primary production

Provisioning Services
Food (crops, livestock, wild foods, etc.)
Fiber (timber, wood fuel)
Genetic resources
Biochemicals, natural medicines, pharmaceuticals
Fresh water

Regulating Services
Air quality regulation
Climate regulation (global, regional, local)
Water regulation
Erosion regulation
Water purification
Pest regulation
Pollination
Natural hazard regulation

Cultural services
Aesthetic values
Spiritual and religious values
Recreation and ecotourism

Source: modified from Patterson, 2009

Fig. 1. Broad categories of mountain ecosystem services

3. On forest multi-functionality

As described in the previous sections, it is commonly known that forests are defined as multi-functional[7] assets, providing, at the same time, different goods, connected with the productive

---

[6] This project, aiming at rationalizing biodiversity conservation in dynamic ecosystems, focuses on assessing the ecological resilience of those components of biological diversity essential for maintaining ecosystem services. It provides this focus in order to suggest priorities for biodiversity conservation policy on the basis of dynamic ecosystems and the services they provide (http://www.rubicode.net).
[7] According to the OECD (2001), the term «multi-functionality» "refers to the fact that an economic activity may have multiple outputs and, by virtue of this, may contribute to several societal objectives at

function (such as timber and non-timber products), and services connected with the ecological functions (such as soil conservation and protection, watercourse protection, hunting, fishing, protection of biodiversity, and the carbon cycle). As presented in the figure 2, these outputs (goods and services) can be classified differently with respect to the parameters of rivalry and excludability (Fisher et al., 2009; Gios & Clauser, 2009; Patterson & Coelho, 2009) and, thus, with reference to forests. These outputs could be purely private (excludable and rival – timber and non timber products) or purely public (non-excludable and non-rival[8] hydro-geological services). Moreover, there is a spectrum of forest goods ranging between purely private and purely public goods. Some of these "intermediate goods" are qualified as club goods (excludable and non-rival – landscape-recreational function) or as open access resources (non-excludable and rival) or "common-pool resources" (Hardin, 1968).). This latter category characterizes many (not just natural) resources. Their need for management has led to the establishment of various institutions in the Alps since the 11th century. Notably, the Autonomous Province of Trento treats almost 60% of overall surface (over 3,000 Km²) land and goods as collective property and manages them as common-pool resources.

Usually, the productive function is defined as the «market function» while the others categories are referred to as «non market functions». In the first case, the forest generates some inputs for the productive processes that can be exchanged in the market and subsequently have a monetary value. Conversely, in the second case, the forest provides public goods (such as carbon sinks) and mixed goods (such as landscape and recreational values) that cannot be exchanged in the market, and therefore, cannot be priced. Moreover, as many studies have demonstrated and as we have already mentioned, "forests have a higher value than that solely connected to production aspects"(Goio et al., 2008).

We should consider that central to the debate on multi-functionality is the degree of the conjunction of the production of secondary goods compared to that of primary goods and the inevitability of this conjunction. Since the 1950s, some authors (Carlson, 1956; Marshall, 1959) have tried to define «joint-production» as involving things that cannot be produced separately, and are joined by a common origin. More recently, according to Shumway et al. (1984), "the joint production encompasses all production situations in which two or more outputs or products are interdependent". These inter-linkages could arise for three different reasons:

1. "there are technical interdependencies in the production process;[9]
2. outputs are produced from a non-allocable input;
3. outputs compete for an (allocable) input that is fixed at the firm level"[10] (OECD, 2001).

---

once. Multi-functionality is, thus, an activity-oriented concept that refers to specific properties of the production process and its multiple outputs."

[8] To briefly clarify, the term «non-rival» means that a unit of the good can be consumed by one individual without diminishing the consumption opportunities available to others, from the same unit. In contrast the term «non-excludable» refers to the situation in which it is physically or institutionally (i.e. through laws) impossible, or very costly, to exclude individuals from consuming a good.

[9] Instead of $Y_1 = f_1 (L_1, K_1, T_1)$ and $Y_2 = f_2 (L_2, K_2, T_2)$, it is $Y_1 = f_1 (Y_2, L_1, K_1, T_1)$ and $Y_2 = f_2 (Y1, L2, K_2, T_2)$. Where:

Y = production
L = labour
K = capital
T = land

This means that, in the case of joint production, the production Y_1 is function not only of the usual production factors (L, K, T) but also of the production Y_2 and vice versa.

In the case of technical complementarities, (1) the products have to be produced together, or, in the other cases (2 and 3), outputs can be produced separately. However joint-production is cheaper because of the presence of economies of scope.

	Excludable	EXCLUDABILITY	Non excludable
Rival	PRIVATE GOODS		COMMON POOL RESOURCES (OPEN ACCESS)
	timber and non-timber products		
	patented processes from genetic resources		local fishing and hunting
			access to genetic materials
RIVALRY	eco-tourism	Hunting and fishing licensing	
	natural parks with entrance fees	water quality trough ecosystem protection	carbon sequestration
	flood control trough ecosystem protection		existence of species and ecosystems
Non rival	CLUB GOODS		PURE PUBLIC GOODS
	Local	LOCALITY	**Global**

Adapted from Landell-Mills & Porras (2002); OECD (2001)

Fig. 2. Different utility flows provided by forests

A second critical aspect is that the time horizon is different depending on the output that is evaluated. The emphasis is usually on the market failures resulting from the difficult assignment of an adequate property rights system.

Within this framework, the multi-functionality aspects that assume greater significance, may be identified as the following: the "type and strength of the link between forest production and secondary products; synergies and trade-offs between the various forest products; specificity of the forest in the provision of services and products not directly commercial; and the fact that the market is unable to assign a price to many secondary products, thereby requiring public intervention" (Henke, 2004). In many cases, as stressed by Janse & Ottitsch (2005), "synergies and the integration of these various components/products is not always without conflict".

---

[10] $Y_1 = f_1 (L_1, K_1, T_1)$ and $Y_2 = f_2 (L_2, K_2, T_2)$ but $L < L_1 + L_2$, $K < K_1 + K_2$, $T < T_1 + T_2$. In this case Y_1 and Y_2 can be produced separately but the costs connected with the production factors are higher than those of the joint production.

Evidently, inappropriate means of development, such as excessive intensification, mechanization, over-exploitation of resources, environmental pollution and urbanization are only some of the factors that could increasingly threaten the multi-functionality of forest ecosystems. Hence, this ecosystem can continue to provide their goods and services, in a rapidly changing world, only if multi-functionality is taken into account in their management. As a consequence, as we discuss in detail in the following sections, with the objective of properly defining the management options and opportunities, it is important to characterize, precisely, the different utility flows performed by forests and to evaluate them.[11] It is clear that the choice of the policy tools that have to be adopted will be: "a) different, depending on whether the goods are private, public or mixed and on the kind of joint production carried out; b) strictly connected with the specific socio-economic and environmental context of a given area; and c) flexible in order to operate within constantly changing circumstances" (OECD, 1999).

4. Some specific reflections on forest join-productions

In the case of forests, we are dealing with the two aspects of joint-production examined in the previous section. In particular, there is technical complementarities with reference to timber production and carbon fixation, and economies of scope with reference to the production of timber and the landscape. As a consequence, the non-commodity outputs are joint products with timber production. This circumstance means that joint products, that clearly create benefits for people living inside and outside the local areas, have different characteristics. In other words while timber is a market product that is "paid", the others are non-market products and are, therefore, "unpaid". It is important to point out the strong impact that the production of non-market outputs, has on the structure of the private costs related to market forest production. In fact, if the increase in private costs exceeds certain levels, this increase may affect the sustainability of the system. In such cases, the most effective solution is to pursue economies of scope rather than scale, because the costs to produce two or more outputs together are less than those for obtaining the same outputs through different production processes.

With the objective of maximizing the environmental externalities associated with forests, the production process should, consequently, be organized in a precise and defined way. The choice of many alpine areas has been and is so now, that of «natural forest management». The problem is that this kind of management causes an increase in the use costs, due to: a) higher cutting and logging costs, related to constraints on the maximum cutting area and to the need to adopt more environmentally friendly techniques.; b) constraints on the characteristics of forest's roads (reduced width, practicable by less efficient equipment); c) the acquisition of heterogeneous material (by species, diameter, features) imposes higher costs for the selection and start-up to the sale.

There are no specific researches able to quantify, in the alpine areas, the additional cost related to the natural forest management. However, some experts (Pollini et al., 1998) estimated that the increase is about 20-30 % of the forest utilisation. In addition, the natural forest management determines also a lower level in the production, and as a consequence, less available wood mass for the next links in the productive chain.

---

[11] The assessment and valuation of ecosystem services, since the seminal papers (Costanza et al., 1997), has "recently focused on an extensive research, with the number of publications increasing almost exponentially" (Fisher et al., 2009).

In particular cases, the cited circumstances lead to the abandonment of the cultivations or to damages to the non-market functions.

In the table 1, we try to link the possible private and social benefits, the different typologies of goods and services provided by forests. This classification occurs, largely, in the alpine context and, in particular, in the area that we are considering as our "case study".

Typology	Private benefits	Social Benefits/Externalities
Forest products	Market based value	Production-chain activities
Hydro-geological	Preservation of forest soil, protection from erosion, landslides, floods	Preservation of lowland soil, water regulation
Climate regulation		Carbon fixation, air depuration
Landscape and recreation	Tourism	Recreational and aesthetic benefits

Table 1. Forest goods and services and private and social benefits

4.1 Brief considerations about markets and payment for ecosystem services

In regard to sustainable forest management,[12] it is necessary to pay particular attention to the costs of multi-functionality and to identify techniques that can internalize the positive externalities provided by forests. This effort will help ensuring a fair distribution of costs and benefits among the local population, the economic actors, the other stakeholders and the entire society. Many authors believe that important opportunities exist for provisioning forest ecosystem services, whether through "governance" (Gibson et al., 2000, 2005), "payment systems and markets" (Engel et al., 2008; Johnson et al., 2001), adjustments to life cycle processes, "or other means " (Patterson & Coelho, 2009). Clearly, as previously mentioned, this process is not easy because of the presence of utility flows that have the characteristics of public or mixed goods.

A category that has been widely analyzed in this context is that of the "Payment for Ecosystem Services" (PES). According to Muradian et al. (2010), the PES[13] "are a transfer of resources between social actors, which aim to create incentives to align individual and/or collective land use decisions with the social interest in the management of natural resources". Wunder (2005) in particular, attributes the following features to the PES:

1.   "a voluntary transaction where,
2.   a well-defined environmental service (or a land use likely to secure that service),

---

[12] During the Second Ministerial Conference on Forest Protection in Europe, held in Helsinki in 1993, the following definition of «sustainable forest management» was introduced: "the correct management and use of forests and forest land in such ways and to such a degree as to conserve their biodiversity, productivity, renewal capacity, vitality and a potential that guarantees their important ecological, economic and social functions both now and in the future, at a local, national and global level without bringing damage to other ecosystems (www.mcpfe.org)". The European Commission (2001), subsequently, stressed that "sustainable forest management is the fundamental aim of development in the forestry sector, where the term «sustainability» refers not only to the regular production of timber, in the forestry sense, but also to the whole range of environmental, economic and social services performed by forests".

[13] Having in mind that "democratic mechanisms for allocating essential and non-substitutable resources may be preferable to markets, at least until basic needs are met" (Farley & Costanza, 2010).

3.   is being 'bought' by a (minimum one) service buyer,
4.   from a (minimum one) service provider,
5.   if and only if the service provider secures service provision (conditionality)."
Very few PES schemes achieve the standards proposed by Wunder (Muradian et al., 2010; Porras et al., 2008). "Generating adequate resources or ensuring a just distribution of payments may require non-voluntary approaches such as taxes or mandatory service charges" (Patterson & Coelho, 2009). Whether payments should be voluntary or coerced through taxation should in fact be determined by the physical characteristics of the resource (Farley et al., 2010; Kemkes et al., 2010). "Services dominated by private good characteristics are amenable to voluntary payments, while services with public good characteristics are not" (Patterson & Coelho, 2009).

5. The landscape-recreational function

In this framework, an example is represented by the last typology presented in the table 1. It is the landscape-recreational function, which we now further analyze.
Natural resources[14] and, thus, forests, under certain conditions[15] that we identify in this and the following sections, could guide the local development of mountain areas, ensuring that income arising from the territory remains with local communities. Although the concept of local development is very broad, according to Greffe (1989, 1990) it can be considered "a process through which a certain number of institutions and/or local people mobilise themselves in a given locality in order to create, reinforce and stabilise activities using, as well as possible, the resources of the territory." In addition, "local development policies can help to achieve sustainable development goals. In fact, they are based on facilitating structural adjustment and enabling economies and societies to adapt to changing conditions, combating social exclusion and maintaining social equilibrium, and making the best use of social, economic and environmental resources in the local area" (OECD, 1999). It should be noted that, the increasing globalization of the economy and changing technologies have opened new markets and new competition with regard to which local development policies need to offer new responses.
According to the paradigm of the total economic value (TEV)[16], which mainly differentiates between use and non-use value, the landscape-recreational function can be subdivided into different components (Table 2). Specifically, "the recreational and scenic values require the direct use of the good: the first one derives from the possibility of carrying out tourist-recreational activities in environmental contests of quality, and the second one is related to the benefits produced by observing certain typologies of landscape" (Goio et al., 2008). In contrast, the evocative value "derives from the desire that a landscape encompassing aesthetic functions should exist, and from knowing that its

---

[14] According to Barbier (2002), "these resources should be viewed as important economic assets, which can be called natural capital".
[15] We are referring, for example, to the control of natural resources, of investments and of legal and administrative rules.
[16] "The concept of total economic value (TEV) is one framework that economists have developed for categorizing the various multiple benefits arising from natural systems" (Barbier, 2002). In particular, it is a tool for the assessment of the intrinsic value of environmental goods aimed at economic evaluation of all functions regardless of their market interest.

associated traditions, culture and lifestyles continue to exist through its conservation" (Novelli, 2005).

From the perspective of local development, each component is related to different management options and to different benefits for people living inside and outside the local area.[17]

Recreational value	Use value
- Areas with user-oriented management[18] - Areas with resources-oriented management[19]	
Scenic value	Use value
Evocative value	Non-use value

Source: Gios & Clauser, 2009

Table 2. Different components of the landscape-recreational function

In table 4 we present some of the possible ways for "internalizing" the landscape-recreational function.

As shown in table 4, these ways are related to the different kinds of goods or services considered. In the case of private goods related to "user-oriented management" the internalization could be a ticket or a fee, while in the case of public goods what is needed is public support. Finally, for mixed goods connected with resources-oriented management, an approach based on the management of "commons" is required.

Typology of goods	Target	Form of internalization
Private	Areas with user-oriented management	Ticket
Public	Landscape as scenery	Public support
Mixed	Areas with resources-oriented management	Approach based on management of "commons"

Table 3. How to internalize the landscape-recreational function

With reference to the landscape-recreational function it is necessary to introduce an element characterizing many mountain areas: the tourism activity. Although this activity can foster the economic development and is a source of employment for the local population, in some cases, it can, also, lead to an imbalance among the various components of ecosystems, producing negative trade-offs (Dollinger, 1988). These trade-offs, sometimes, become very difficult to manage.

---

[17] "One particular landscape typically has different functions for different people" (Heilig, 2003).

[18] We refer to areas that generate direct revenue. These include the following:
- areas with quick and excludable admission (e.g. adventure parks and golf courses) (Type I) and
- areas characterised by the provision of direct use[18] services and ad hoc facilities accessible through the payment of fees (e.g. hunting, fishing and mushroom collection) (Type II).

[19] Includes areas characterized by the provision of direct use, free of charge services (that is, open-access protected areas) that, under certain conditions, allow the creation of other sources of revenue (e.g. restaurants, hotels and guides) (Type III).

6. Some policy implications for decision-makers

In this framework, if the «ultimate aim» is the enhancement of the landscape-recreational function, a strategy able to incorporate jointly, forest management, the kind of landscape-recreational components and, finally, the characteristics of the tourism system has to be adopted. For this purpose, it is really important to take into account not only the specific characteristics of the forests, but also the system in which they are included. These considerations are summarized in the following table (n°4).

When the landscape is referred to as a specific resource for a well-defined project (for example an adventure park), the arrangements for tourist activities require large investments in equipment and structures with related management costs. In the area under consideration, cash flow can create, both directly and indirectly, jobs and sources of income. It also represents the underpinning of the traditional development pattern of some touristic districts, which has occurred since the 1960s in alpine areas. In other words, investments transform a public good into a private one. In contrast, in cases where the investments needed to utilize the natural resources are of small dimension, it is impractical to implement mechanisms of excludability from consumption, even if such mechanisms were technically feasible. It has to be noted that the forms of tourism, so-called "green" or "soft", fall mainly into this category.

Forest Landscape as:	Touristic system	Forest utilization	Type of intervention
1) Specific resource (adventure park[20])	Specific touristic project	Specialize areas oriented to a prevalent use	Active: equipment investments
2) Scenery	Weak and uni-directional links with the touristic system	maximization of biomass	Passive: diminishing the utilization, check fire and pest
3) "Complex" resource (visit to natural park)	Strong and bi-directional links with the touristic system	Naturalistic selvi-culture	Active: knowledge and dissemination investments

Table 4. Intervention related to tourism exploitation of forests

The central objective is to find, even in the case of landscape and recreational activities that do not require large investments, mechanisms that allow the enjoyment of those activities after a specific payment is made as a compensation supporting local development.[21]

---

[20] They are acrobatic paths realized in forested areas that allow direct contact with nature and the possibility of directly exploiting the trunks of the trees for the preparation of the various paths. These paths are very well developed in France, the United Kingdom (www.ttadventure.co.uk) and Italy (www.agilityforest.it).

[21] A good example, in this context can be represented by the "Sentiero del Castagno" (Alto Adige, Italy, http://www.valleisarco.info/it/attivita/estate/escursionismo/sentiero-del-castagno.html) or "Les Route du Bois in Belgium, (www.lesroutesdubois.be) .

In addition some further observation can be made. In the case of specific resources, the forest is managed to extract timber. However, this benefit is a "sub-product" because the dominant use of the forest is as an adventure-park. In the second case, where the forest is considered as scenery (for example in relation to sport use), an increase in the overall timber quantity produced, is a positive aspect. Its perceived value in fact increases, but the associated management costs, which clearly grow (for example for checking fires and pests), are not compensated by the market. Consequently a public support is needed. Finally, with respect to "complex resources" a double objective, related to tourism activity and timber production, has to be achieved. In this case, higher costs related to timber production should be compensated by tourism revenues. However that initiative requires investments to disseminate and share knowledge amongst the general stakeholders who use natural resources.

6.1 The need for a participative approach

The future development, especially of mixed goods in the Alps, will depend, largely, on the ability to involve local stakeholders in the environmental protection and promotion processes, establishing, at the same time, the priorities for each single area. Clearly, to make this involvement work, local actors should be able to direct the management of natural resources, in general, and of forests in particular, towards their own interests and needs. They also need to control the management options adopted.[22] As known, the participative approach of local stakeholders has emerged in the last 15 years, following, precisely, the evolution of the concept of sustainable development. It is based on the belief that citizens are able to shape their own future. It is "thought up on the conviction that people are capable of defining their own future" (Jennings, 2000). As a consequence, "it uses capabilities and local knowledge to guide and define the nature of actions and strategies" (Jennings, 2000). Through efficient participative development processes, it is possible to take into account some territorial dimensions that are often neglected or not considered, such as traditions, beliefs and habits, thereby creating the preconditions for the implementation of spontaneous action by the communities involved. In the development initiatives related to the management of natural resources, however, this participation cannot be confined to the mere application of techniques to facilitate the involvement of larger social groups. On the contrary, it is fundamental that the stakeholders become aware of the issues related to natural, social and human capital, thus creating a shared sense of the problems and the basis for potential collective actions. Since sustainability refers to three different dimensions (environmental, social and economic), integrated and comprehensive territorial development is required. In order to pursue it, as already noted, it is necessary to understand how the stakeholders perceive the factors connected with sustainability, facilitating and promoting an enriching exchange of views, knowledge and initiatives. To obtain a complete and reliable overview, these exchanges should involve both public and private actors, producing new synergies and new partnerships in the area.

This field is very complex and a pre-eminent participative process does not exist, because the process "is perceived and implemented in different ways" (Buchy & Hoverman, 2000). In fact

---

[22] According to Carpenter and Folke (2006) "management actions should be viewed as experiments that can improve knowledge of social–ecological dynamics if the outcome is monitored and appropriately analyzed".

in the literature there are many different classifications of participation[23], "because of the concept give rise to a wide range of interpretations" (Lawrence, 2006). Some writers take into account the degree of involvement, which can be strong or weak. For the World Bank (1996), in fact, "participation is strong if there is a real influence on development decisions by local actors and weak in the case of a simple informative involvement concerning the implementation or benefits of a particular development activity". Other classifications (Rowe & Frewer, 2000) focus on the nature rather than the degree of engagement, identifying different types of public engagement by the direction of communication flows between parties. According to this view, information dissemination to passive recipients constitutes "communication", gathering information from participants is "consultation" and "participation" is conceptualized as two-way communication between participants and exercise organizers in which information is exchanged in some sort of dialogue or negotiation. Others (Biggs, 1989) describe the level of engagement as a relationship that can be "contractual", "consultative", "collaborative" and "collegiate". Finally, some engagements stand between pragmatic participation and normative participation. "The first focuses on process, suggesting that people have a democratic right to participate in environmental decision-making", while the second "arguments focus on participation as a means to an end, which can deliver higher quality decisions " (Reed, 2008).

For this reason, different kinds of participation can be implemented, in relation to: "a) the characteristics and conditions of any specific context, b) the aims that have to be realized, and c) the ability of the stakeholders to influence the final results" (Richards et al., 2004; Tippett et al., 2007). The literature urges to move towards a high degree of participation (Arnstein, 1969; Johnson et al., 2004) or to a strong participation, as defined by the World Bank (1996). In addition, several authors believe that the success of participatory processes should be institutionalized[24] (Richards et al. 2004). They require a) "a sufficiently detailed and clear description of the context and objectives" (Reed, 2008), b) "to identify appropriately and adequately the role of each actor, be it public or private "(Purnomo et al., 2005), to manage any conflicts, and c) to encourage the actors to develop an adequate motivation and ability to participate, triggering a process that might be called «educational»[25]. This approach requires

---

[23] For example:

- "Participation is concerned with . . . the organised efforts to increase control over resources and regulative institutions in given social situations on the part of groups and movements hitherto excluded from such control" (Pearse & Stiefel, 1979);

- "Participation can be seen as a process of empowerment for the deprived and the excluded. This view is based on the recognition of differences in political and economic power among different social groups and classes. Participation in this sense necessitates the creation of organisations of the poor which are democratic, independent and self- reliant!" (Ghai, 1988);

- "Participation is a process through which stakeholders influence and share control over development initiatives and the decisions and resources which affect them" (World Bank, 1996);

- "Participatory development stands for partnership which is built upon the basis of dialogue among the various actors, during which the agenda is jointly set, and local views and indigenous knowledge are deliberately sought and respected. This implies negotiation rather than the dominance of an externally set project agenda. Thus people become actors instead of being beneficiaries" (OECD, 1994).

[24] If participation is a democratic right and not just a legislative goal.

[25] The issue is really one of education and politics: neither the general public nor decision makers appear to be well-informed concerning the relative contributions of ecosystem services and economic growth to our well-being" (Farley et al., 2010).

knowledge of each actor to check the degree of understanding and awareness of the project, with the aim of filling any gaps. Today, it is increasingly necessary to include in the participatory processes the so-called «experts»: a) "local stakeholders, namely those who have proven experience and knowledge (location-specific), not only scientific but also operational, in reference to the area and location of interest, b) the external stakeholders, namely those who truly understand the phenomena and who have a more scientific/universal knowkledge" (McCall, 2003; Fraser & Lepofsky, 2004). These categories are generally involved "simultaneously" (Reed, 2007), because local knowledge combined with what science can contribute, leads to a more complete understanding of complex systems and processes (Johnson et al. 2004), as well as the learning pathways within each category and between the two categories.

6.2 Evidence from alpine examples: Logarska Dolina

As presented in table 4, specific economic and political tools need to be created in order for stakeholders to pursue local development. Payment for ecosystem services is rather common for some resources, but not for scenery and landscape services.

There are examples of fees connected with the touristic and recreational uses of alpine territories, but they are mainly not resource-related. Many local communities are requiring daily-payment from tourists, but they cannot be individuated as in the categories of Table 4 because their practices have general purposes and, sometimes, are not supportive of the maintenance of landscape. In fact, it is necessary to aim for payments capable of financing management activities devoted to preserving and enhancing the natural and forest features of the Alps.

The role of the participation of local inhabitants is clearly presented with reference to "Logarska Dolina": a valley seven kilometers long, covered by meadows and forests", with some waterfalls, is located in the northern part of Slovenia. The attraction of this area lies in its abundant natural sights, coupled with an almost pristine environment. It has been attracting hikers since the late 1800s. Also characteristic are the farmsteads, which have, over the centuries, aided in building a cultural landscape. This valley has recently been part of the network of the "European Destination of Excellence".[26]

In 1987, the local council of Mozirje decided to create a "Landscape Park". However, there were not enough financial resources for protecting the local flora and fauna and developing the recreational and management structures needed. An additional problem has been the increasing number of tourists visiting the valley by car and creating such diseconomies as pollution, chaos, fires and rubbish.

The small local population (at that time, ten families with 35 persons) decided to prevent an excessive use of the local amenities. According to local oral norms, the land is private and it has not been divided over the centuries through inheritance. The only public property is the

---

[26] The population of this district is primarily engaged in forestry, animal husbandry and, most recently, tourism, the prosperity from which is largely supported by this area's great natural beauty. An unspoiled natural environment, coupled with the fact that this region had not been overdeveloped, has worked to the advantage of the local community. However, the people of the Solcava District are well aware that this pristine environment must be preserved at all costs. For this reason, they have chosen to develop high quality tourism, which emphasizes the individual, offering him peace as well as the opportunity to enjoy an active holiday in harmony with nature" (from official website www.logarska-dolina.si).

traffic road situated in the middle of the valley. Although some families were able to make a living through their occupations in agriculture and forestry, others might be interested in developing some services for tourists. The local population agreed that the option of direct management was the optimal choice, even though it involved delegating this activity to external bodies. In 1992, the council of Mozirje created a public company devoted to the management of the park of the Logarska Dolina valley. There are 14 members: ten local families, two companies owning houses in the valley for the recreation of their staff, the manager of the local hotel and a member of the local tourist board. Thus, decisions are made only by persons who are living in the valley or who have a strong interest in the valley. They are members who are aiming at long-term results by taking care of the cultural landscapes as well as the employment of locals.

A decision to introduce a fee for visiting the valley was made to reduce the number of cars and to restore a sense of quiet. The valley should be crossed, preferably, by foot, with tourists leaving their cars at the free parking places located on private land at the beginning of the valley. The entering fees generated most of the money needed to manage the public company. However, more recently, most financial support has been coming from external sources, such as those from the European Union or national government. The small managing board permitted a review of a sustainable model of development, which rejected the ideas of creating golf courses, tennis halls, or new buildings on land not utilized. Few developments were permitted other than nearby local farms in order to avoid landscape fragmentation. "We are creating our house, not a leisure park," reflects the main aim of the board of the public company. What has been created is a heating network, using biomasses, and a purification plant. Five permanent positions have been created for managing the local resources and the activity of the public company.

In 2008, the Logarska Dolina Valley was incorporated into a large park that included the surrounding three valleys of the Solcava district. The cooperative system of management was used as an example. However, it has not been incorporated into the new authority. "The sense of trust and reciprocity is easier when there are few persons involved and all of them are sharing almost the same interests; the largest and most direct right of participation is the better way of management" declared Avgust Lenar, the Director of the Logarska Dolina public company (CIPRA, 2007).

Similar communities, which required the payment of a fee for entering the territories, could be found within the Alps in Natural Parks or in some municipalities where some outstanding recreational or landscape features are located (Cimoliana Valley in Friuli, Italy, nearby the peaks of Cime di Lavaredo).

Dimensioning the population involved, the property regimes and the structure of the political and management organization is playing a relevant role in defining the participatory tools and the design of the development plan and strategies. As previously mentioned in Paragraph 6.1, there are no models that can easily be adapted for the Alps. Thus, there is need to adapt or create new participatory tools and strategies, paying particular attention to the characteristics of territory and stakeholders.

7. Acknowledgment

The authors would like to take this opportunity to acknowledge the Autonomous Province of Trento (PAT) for providing financial support for the research, carried out within the activities of the project "Public policies and local development: innovation policy and its effects on locally embedded global dynamics (OPENLOC)" (2008-2012; www.openloc.eu).

8. References

Arnstein, A. (1969). A ladder of citizenship participation. Journal of the American Institute of Planners, vol. 26, pp. 216-233.

Barbier, E.D. (2002). *The Role of Natural Resources in Economic Development.* CIES Discussion Paper 0227.

Biggs, S. (1989). Resource-Poor Farmer Participation in Research: a Synthesis of Experiences From Nine National Agricultural Research Systems. OFCOR Comparative Study Paper, vol. 3. International Service for National Agricultural Research, The Hague

Brundtland, G. (ed) (1987). *Our Common future.* The World Commission on Environment and Development. Oxford, Oxford University Press.

Buchy, M., Hoverman, S. (2000). Understanding Public Participation in Forest Planning: a Review. *Forest Policy and Economics,* vol. 1, pp. 15-25.

Carlson, S. (1956). *A study on the pure theory of production.* Kelley and Millman, New York.

Carpenter, S.R., Folke, C. (2006). Ecology for transformation. *Trends in Ecology & Evolution,* vol. 21, pp. 309-315

CIPRA, (2007). Implementing knowledge – making use of local potentials. In: CIPRA-Info 82, March 2007, pp. 26-30.

Costanza, R., d'Arge, R., de Groot, R., Farberk, S., Grasso, M., Hannon, B., Limburg, K., Naeem, S., O'Neil, R.V., Paruelo, J., Raskin, R.G., Suttonkk, P., van den Belt, M. (1997). The value of the world's ecosystem services and natural capital. *Nature* vol. 387, pp. 253-260.

Daily, G.C., Alexander, S., Ehrlich, P.R., Goulder, L., Lubchenco, J., Matson, P.A., Mooney, H.A., Postel, S., Schneider, S.H., Tilman, D. and Woodwell, G.M. (1997). Ecosystem services: benefits supplied to human society by natural ecosystems. *Issues in Ecology* vol. 2. Ecological Society of America, Washington D.C.

Dollinger F. (1988). Die Salzburger Naturraumpotentialkartierung. Theoretische Grundlagen des Projektes aus der Sicht des Naturraumpotentialkonzeptes und Ableitung von Bearbeitungsrichtlinien. Mitteilungen und Berichte des Salzburger Institutes für Raumforschung, 3+4/1988.

Engel, S., Pagiola, S., Wunder, S. (2008). Designing payments for environmental services in theory and practice: an overview of the issues. *Ecological Economics,* vol. 65, pp. 663-674.

European Community Commission (2001). Ambiente 2010: il nostro futuro, la nostra scelta. Sesto programma di azione per l'ambiente. *Comunicazione della Commissione al Consiglio, al parlamento Europeo, al Comitato Economico e Sociale e al Comitato delle Regioni. COM (2001) 31 definitivo,* Bruxelles, 24.1.2001.

FAO (2005). Third expert meeting on harmonizing forest-related definitions for use by various stakeholders, *Proceedings* FAO, Rome, 17-19 January 2005

Farley, J., Aquino A., Daniels A., Moulaert A., Lee D., Krause A. (2010). Global mechanisms for sustaining and enhancing PES schemes. *Ecological Economics,* vol. 69, pp. 2075-2084.

Farley J, Costanza R. (2010). Payments for ecosystem services: From local to global. *Ecological Economics,* vol. 69, pp. 2060-2068.

Fisher, B., Turner, R.K., Morling, P. (2009). Defining and classifying ecosystem services for decision making. *Ecological Economics,* vol. 68, pp. 643-653.

Fraser, J., Lepofsky, J. (2004). The use of knowledge in neighborhood revitalization. *Community Dev. J.,* vol. 39 (1), pp. 4-12.

Ghai, D. (1988). Participatory Development: Some Perspectives from Grassroots Experiences'. Discussion Paper No. 5., UNRISD Geneva.

Gibson, C., McKean, M.A., Ostrom, E. (2000). *People and Forests: Communities, Institutions, and Governance.* MIT Press, Cambridge.

Gibson, C., Ostrom, E., Williams, J.T. (2005). Local enforcement and better forests. *World Development*, vol. 2, pp. 273–284.

Gios, G., Clauser, O. (2009). Forest and tourism: economic evaluation and management features under sustainable multi-functionality. *iForest* 2, pp. 192-197. Article available online at: http://www.sisef.it/iforest/

Goio, I., Gios, G., Pollini, C. (2008). The development of forest accounting in the province of Trento (Italy). *Journal of Forest Economics*, vol. 14, pp. 177–196.

Greffe, X. (1989). *Decentraliser pour l'Emploi. Les Initiatives Locales de Développement.* Economica. Paris

Greffe X. (1990). Le Développement Économique Local, *Commissione Europea DGV*, Bruxelles.

Handoh, I.C., Hidaka T. (2010). On the timescales of sustainability and futurability. *Futures* vol. 42, pp. 743–748

Hardin, G. (1968). The Tragedy of the Commons. *Science*, 162 (3859), pp. 1243-1248.

Heilig, G.K. (2002). Multifunctionality of Landscapes and Ecosystem Services with Respect to Rural Development. In *Sustainable Development of Multifunctional Landscapes*, Helming, K., Wiggering, H. (Eds.). Berlin, New York (Springer Verlag)

Henke, R. (2004). *Verso il riconoscimento di una agricoltura multifunzionale. Teorie, politiche, strumenti.* Edizioni Scientifiche Italiane, Roma, Italy.

Janse, G., Ottitsch, A. (2005). Factors influencing the role of non-wood forest products and services. *Forest Policy and Economics*, 7, pp. 309–319.

Jennings, R. (2000). Participatory Development as New Paradigm: the Transition of Development Professionalism, Prepared for the "Community Based Reintegration and Rehabilitation in Post-Conflict Settings" Conference Washington, D.C.

Johnson, N., White, A., Perrot-Maitre, D. (2001). Developing Markets for Water Services From Forests: Issues and Lessons for Innovators. Forest Trends, World Resources Institute, and The Katoomba Group, Washington, DC.

Johnson, N., Lilja, N., Ashby, J.A., Garcia, J.A. (2004). Practice of participatory research and gender analysis in natural resource management. *Natural Resources Forum,* vol. 28, pp. 189–200.

Kemkes, R. J., Farley J., Koliba C. J. (2010). Determining when payments are an effective policy approach to ecosystem service provision. *Ecological Economics*, vol. 69, pp. 2069-2074

Landell-Mills, N., Porras, I. (2002). *Silver bullet or fool's gold? A global review of markets for forest environmental services and their impacts on the poor.* Instruments for sustainable private sector forestry series, IIED, London, UK.

Lange, G.M. (2004). Manual for Environmental and Economic Account for forestry: a tool for cross-sectoral policy analysis. *Working Paper*, FAO, Forestry Department, Rome, Italy.

Lawrence, A., 2006. No personal motive? Volunteers, biodiversity, and the false dichotomies of participation. *Ethics, Place and Environment*, vol. 9, pp. 279–298.

Mantau, U., Mertens, B., Welcker, B., Malzburg, B. (2001). Risks and chances to market recreational and environmental goods and services -- experience from 100 case studies. *Forest Policy and Economics*, vol. 3, pp. 45-53.

Mantau, U., Wong, J.L.G., Curl, S., 2007. Towards a Taxonomy of Forest Goods and Services. *Small-Scale Forestry*, vol. 6, pp. 391-409.

Marshall, A. (1959). *Principles of economics*, Macmillan, London.

McCall, M.K. (2003). Seeking good governance in participatory-GIS: a review of process and governance dimensions in applying GIS to participatory spatial planning. *Habitat Int.*, vol. 27, pp. 549–573.

Millennium Ecosystem Assessment (MEA) Reports, (2005). Ecosystems and Human Well-being: Current State and Trends. Chapter 24 "Mountain Systems". Island Press, Washington, USA.

Muradian, R., Corbera, E., Pascual, U., Kosoy, N., May, P.H. (2010). Reconciling theory and practice: an alternative conceptual framework for understanding payments for environmental services. *Ecological Economics*, vol. 69, pp. 1202–1208.

Newman, L. (2005). Uncertainty, innovation, and dynamic sustainable development, *Sustain. Sci. Pract. Policy*, vol. 1, pp. 25–31.

Novelli S., 2005. Aspetti economici e politici della conservazione del paesaggio rurale. Definizione delle strumento di indagine per una valutazione economica nell'astigiano. Tesi di dottorato ciclo XVI. Facoltà di Agraria, Università degli Studi di Torino.

OECD, (1994). *Promoting Participatory Development*, Paris.

OECD, (1999). *Best Practices in Local Development*. LEED, Notebook, 27, Paris.

OECD, (2001). *Multi-functionality. Towards an Analytical Framework*. Paris.

OECD, (2008). *Strategic Environmental Assessment and Ecosystem Services*, Paris.

Patterson, T.M., Coelho, D.L. (2009). Ecosystem services: Foundations, opportunities, and challenges for the forest products sector. *Forest Ecology and Management*, vol. 257, pp. 1637–1646

Pearse A., Stiefel M. (1979). *Inquiry into Participation*, UNRISD, Geneva

Peters, C.M., Gentry, A.H., Mendelsohn, R. (1989). Valuation of an Amazonian Rainforest. *Nature*, vol. 339, pp. 655-656.

Pollini, C., Spinelli, R., Tosi, V. (1998). *Tecniche per una gestione multifunzionale durevole dei boschi della montagna alpina: l'esperienza del progetto LIFE in Trentino*. Comunicazione di ricerca 98/1 I.T.L. C.N.R. S. Michele a/A (Tn), Grafiche Artigianelli, Trento

Porras, I., Grieg-Gran, M., Neves, N. (2008). *All That glitters: A Review of Payments for Watershed Services in Developing Countries*. The International Institute for Environment and Development, London.

Purnomo, H., Mendoza, G.A., Prabhu, R., Yasmi, Y., (2005). Developing Multi-Stakeholder Forest Management scenarios: a Multi-Agent System Simulation Approach Applied in Indonesia. *Forest Policy and Economics*, vol. 7, pp. 475-491.

Reed, M.S. (2007). Participatory technology development for agroforestry extension: an innovation-decision approach. *African Journal of Agricultural Research*, vol. 2, pp. 334–341.

Reed, M.S. (2008). Stakeholder participation for environmental management: A literature review. *Biological Conservation*, n° 141 (10), 2417 –2431

Richards, C., Sherlock, K., Carter, C. (2004). Practical Approaches to Participation. Socio-Economic Research Programme (SERP). The Macaulay Institute, Aberdeen.

Rowe, G., Frewer, L. (2000). Public participation methods: a framework for evaluation in science. *Technology and Human Values*, vol. 25, pp. 3–29.

Shumway, C.R., Pope, R.D., Nash, E.K. (1984). Allocable fixed inputs and jointness in agricultural production: implications for economic modeling", American Journal of Agricultural Economics, n° 66(1), pp. 72-78.

Tippett J., Handley J.F., Ravetz J., 2007. Meeting the challenges of sustainable development – A conceptual appraisal of a new methodology for participatory ecological planning. *Progress in Planning*, vol. 67, pp. 9–98.

World Bank, (1996). *The World Bank Participation Sourcebook*. Washington D.C..

Wunder, S. (2005). Payments for Environmental Services: Some Nuts and Bolts. Occasional Paper No. 42. Center for International Forestry Research, Nairobi, Kenya.

4

Implementation of the U.S. Legal, Institutional, and Economic Criterion and Indicators for the 2010 Montreal Process for Sustainable Forest Management

Frederick Cubbage[1], Kathleen McGinley[2], Steverson Moffat[2],
Liwei Lin[1] and Guy Robertson[2]
[1]North Carolina State University Department of Forestry and Environmental Resources
[2]U.S. Department of Agriculture, Forest Service
USA

1. Introduction

At the 1992 United Nations "Earth Summit" in Rio de Janeiro, most of the countries in the world, including the United States, agreed to international accords to protect biodiversity and mitigate climate change. However, they could not agree on a convention for forests, because developing countries wanted to preserve their autonomy and sovereign control of their forest resources, and developed countries would not guarantee them financial support to protect their forests (Humpheys 2006). This failure eventually led to the development of multi-lateral forest agreements and treaties to at least measure and monitor forest sustainability through Sustainable Forest Management Criteria and Indicators (SFM C&I), as well as the movement to create forest certification programs for sustainable forestry.

The creation of multilateral SFM C&I frameworks were a public response to the lack of a binding international agreement on forests; similarly, the development of forest certification systems were a non-state market driven response (Cashore et al. 2004). SFM C&I processes have since been developed to measure and monitor various conditions of forest sustainability at the national or regional level. Forest certification, on the other hand, was developed to also measure SFM, but at the forest management unit level. Many efforts have been made to harmonize national-level SFM C&I with national forest certification efforts, particularly in Europe.

These various efforts at measuring, monitoring, and encouraging SFM address biophysical, economic, and social aspects of forest systems. Many of the C&I efforts have made considerable progress at tracking biophysical characteristics of forests, but the measurement and monitoring of legal and institutional features has developed more slowly. Furthermore, determining whether we are achieving SFM, in general, and if our laws and institutions are helping, in particular, is difficult to ascertain.

In this book chapter, we discuss the development of one criterion of SFM C&I in the United States—the Legal, Institutional, and Economic Criterion and Indicators for the 2010 Montreal Process for Sustainable Forest Management (Criterion 7). This criterion has the greatest number of indicators of the seven Criteria developed by its participating countries, yet most of these are not easily measured or tracked. Thus, this paper describes the approach that we developed in the United States to measure and discuss the legal and institutional indicators for SFM. Criterion 7 and its Indicators have been revised since the U.S. National Report on Sustainable Forests (USDA Forest Service 2011) was issued, and those revisions and suggestions for the next round of C&I reporting also are discussed.

2. International agreements to measure, monitor, and report on SFM

The International Tropical Timber Organization (ITTO) is considered the pioneer of international C&I development, publishing its first framework of C&I for tropical forests in 1992 (Humphreys 2006). That same year, at the United Nations Conference on the Environment and Development (UNCED) in Rio de Janeiro, the non-binding plan of action known as "Agenda 21" and Statement of Forest Principles were signed by more than 178 countries (www.un.org/esa/dsd/agenda21/). These non-binding agreements included a call for the development of international criteria for monitoring national forest resources in all forest types (McDermott et al. 2010).

This combination of the initial ITTO C&I work and the UNCED agreements led to the development of eight regional C&I processes—African Timber Organization, Asia Dry Forests, Dry-Zone Africa, Lepaterique (Central America), Montreal (Non-European Temperate and Boreal), Near East, Pan-European Forest, and Tarapoto (Amazon). The Montreal and Pan-European (now known as the Ministerial Conference on the Protection of Forests in Europe (MCPFE)) Processes were the first to develop C&I frameworks in the mid-1990s, adopting comparable sets of national level C&I for the sustainable management of temperate and boreal forests (The Montreal Process 2009). Today, more than 150 countries are engaged in one or more regional and/or international SFM C&I process (Wijewardana 2008).

As of 2011, the Montreal Process includes 12 member countries—Argentina, Australia, Canada, Chile, China, Japan, Republic of Korea, Mexico, New Zealand, Russian Federation, United States of America, and Uruguay. The multilateral Montreal Process demonstrates that the countries agree on the importance of improving understanding of and measuring progress toward SFM (www.mpci.org). The Montreal Process framework of Criteria and Indicators for the Conservation and Sustainable Management of Temperate and Boreal Forests was adopted initially through the Santiago Declaration in 1995. This covered 7 Criteria and 67 associated specific Indicators. Criteria reflect broad principles or themes that measure forest sustainability; while specific Indicators can be used to determine whether these principles are being achieved. As a whole, the C&I framework serves as a tool for assessing trends in forest condition and management at the national level and as a common framework among countries for describing, monitoring and evaluating progress towards sustainability at both national and international levels (The Montreal Process 1999). This framework has also grown to serve as a standard reference for many national statistics about forests in the U.S., both in the National Report on Sustainable Forests and in separate supplemental reports and web based data bases.

The initial Montreal Process Criteria for forest conservation and management were intended to measure and monitor forest sustainability with the best indicators possible. Sustainability generally refers to the classic 1987 Brundtland Report definition to "provide for the needs of the present generation without compromising the ability of future generations to meet their needs." This definition of sustainability has evolved to include ecological, economic, and

Implementation of the U.S. Legal, Institutional, and Economic Criterion and Indicators
for the 2010 Montreal Process for Sustainable Forest Management

51

social components. The Montreal Process drew from these principles to develop broad criteria that are listed below:

Criterion 1: Conservation of biological diversity

Criterion 2: Maintenance of productive capacity of forest ecosystems

Criterion 3: Maintenance of forest ecosystem health and vitality

Criterion 4: Conservation and maintenance of soil and water resources

Criterion 5: Maintenance of forest contribution to global carbon cycles

Criterion 6: Maintenance and enhancement of long-term multiple socio-economic benefits to meet the needs of societies

Criterion 7: Legal, institutional and economic framework for forest conservation and sustainable management

In general, each Montreal Process member country develops its own approach to measuring and monitoring Indicators, although the Montreal Process Working and Technical Groups facilitate discussions among members and provide technical guidance. In 1997, Montreal Process member countries produced an Approximation Report that provided information on the status of data availability and collection with emphasis on significant implementation issues related to the C&I. The first national reports on the 7 Criteria and 67 Indicators were released in 2003 by participating member countries. These reports varied in the extent and depth to which they covered the suite of C&I. Overall, the 2003 efforts revealed that most countries regularly collected most of the data needed to report conditions with regards to SFM biophysical Indicators, but struggled to address the largely qualitative economic, social, and institutional Indicators in Criterion 7.

Subsequently, the Montreal Process Working Group initiated a process to revise the original C&I, based on experiences with their implementation. At a Montreal Process meeting in Buenos Aires, Argentina in 2007, member countries agreed to revisions of the Indicators associated with the first six Criteria. These Criteria were retained as originally proposed, but some of the Indicators were changed or deleted and new Indicators were added.

In a 2009 meeting in Korea, member countries agreed to revisions of Criterion 7 and its Indicators, including a change to the title of the criterion to "Legal, policy, and institutional framework", as well as a decrease from 20 to 10 Indicators. For the 2010 reporting cycle, member countries had time to incorporate the revised Indicators for Criteria 1 – 6, but the modified Indicators under Criterion 7 were released too late to be analyzed and integrated with the 2010 country reports. Table 1 summarizes the original and revised Indicators under Criterion 7. The revised 2010 Montreal Process reports had 64 Indicators, and with no further changes, the 2015 reports will have 54 Indicators.

3. Criterion 7 developments

Criterion 7 and its original 20 Indicators are intended to address the crucial question of whether current laws, institutions, and economic structures are adequate to sustainably manage and conserve a nation's forests. The importance of the legal, institutional and economic framework in forest conservation and sustainable management to the Montreal Process participants is clear given the quantity and breadth of the original Indicators. Most of these Indicators, however, are not amenable to concise quantified measurement. Characterizing national trade policies in terms of their impact on forest sustainability (Indicator 7.3.b), for example, entails an analysis framework and synthesis of information at the level of a full research paper. However, Indicator 7.3.b is but one of 20 Indicators under Criterion 7, and one of 64 within the entire suite of C&I in the 2010 reports.

Table: Initial Criterion 7 Indicators, 1995-2010	Table: Revised Criterion 7 Indicators, 2011+
Criterion 7: Legal, Institutional and Economic Framework for Forest Conservation and Sustainable Management	*Criterion 7: Legal, Policy, and Institutional Framework*
7.1 Legal and Policy Framework	
7.1 Extent to which the legal framework (laws, regulations, guidelines) supports the conservation and sustainable management of forests, including the extent to which it:	7.1.a Legislation and policies supporting the sustainable management of forests.
7.1.a Clarifies property rights, provides for appropriate land tenure arrangements, recognizes customary and traditional rights of indigenous people, and provides means of resolving property disputes by due process;	7.1.b. Cross-sectoral policy and programme coordination
7.1.b Provides for periodic forest-related planning, assessment, and policy review that recognizes the range of forest values, including coordination with relevant sectors;	
7.1.c Provides opportunities for public participation in public policy and decision-making related to forests and public access to information;	
7.1.d Encourages best practice codes for forest management;	
7.1.e Provides for the management of forests to conserve special environmental, cultural, social and/or scientific values.	
7.2 Extent to which the institutional framework supports the conservation and sustainable management of forests, including the capacity to:	7.2.a Taxation and other economic strategies that affect the sustainable management of forests.
7.2.a Provide for public involvement activities and public education, awareness and extension programs, and make available forest-related information;	
7.2.b Undertake and implement periodic forest-related planning, assessment, and policy review including cross-sectoral planning and coordination;	
7.2.c Develop and maintain human resource skills across relevant disciplines;	
7.2.d Develop and maintain efficient physical infrastructure to facilitate the supply of forest products and services and support forest management;	
7.2.e Enforce laws, regulations and guidelines	
7.3 Extent to which the economic framework	7.3a Clarity and security of land

(economic policies and measures) supports the conservation and sustainable management of forests through:	and resource tenure and property rights
7.3.a Investment and taxation policies and a regulatory environment which recognize the long-term nature of investments and permit the flow of capital in and out of the forest sector in response to market signals, non-market economic valuations, and public policy decisions in order to meet long-term demands for forest products and services;	7.3.b Enforcement of laws related to forests
7.3.b Non-discriminatory trade policies for forest products	
7.4 Capacity to measure and monitor changes in the conservation and sustainable management of forests, including:	7.4.a Programmes, services, and other resources supporting the sustainable management of forests
7.4.a Availability and extent of up-to-date data, statistics and other information important to measuring or describing indicators associated with criteria 1-7;	7.4.b Development and application of research and technologies for sustainable management of forests
7.4.b Scope, frequency and statistical reliability of forest inventories, assessments, monitoring and other relevant information;	
7.4.c Compatibility with other countries in measuring, monitoring and reporting on indicators	
7.5 Capacity to conduct and apply research and development aimed at improving forest management and delivery of forest goods and services, including:	7.5.a Partnerships to support the sustainable management of forests
7.5.a Development of scientific understanding of forest ecosystem characteristics and functions;	7.5.b Public participation and conflict resolution in forest-related decision making
7.5.b Development of methodologies to measure and integrate environmental and social costs and benefits into markets and public policies, and to reflect forest-related resource depletion or replenishment in national accounting systems;	7.5.c Monitoring, assessment and reporting on progress towards sustainable management of forests
7.5.c New technologies and the capacity to assess the socio-economic consequences associated with the introduction of new technologies;	
7.5.d Enhancement of ability to predict impacts of human intervention on forests;	
7.5.e Ability to predict impacts on forests of possible climate change	

Table 1. Initial and Revised Indicators for Montreal Process Criterion 7: Legal, Institutional and Economic Framework for Forest Conservation and Sustainable Management

The time, financial, and human resources available for the development of each Indicator are limited, as is space for reporting. Moreover, the US National Report on Sustainable Forests is set up to provide concise two-page reports on the importance, status and change in each Indicator, albeit longer technical reports for each Indicator are available in an on-line database. This is not just a matter of limited space for analysis, but also reflects the broad scope for different levels of details and perspectives in the analysis. Comparing the data and numbers in the comparable two-page summaries within and between reports is much easier than comparing two larger associated research papers.

Much of the Indicator development for Criterion 7 in the 2003 National Report relied on separate narrative assessments that identify key concepts and policy components, but which are not regularly collected or monitored and are difficult to update in a consistent fashion. Other Montreal Process Working Group countries had similar results from their efforts to address Criterion 7, largely resulting in revisions of these Indicators to a more qualitative structure. Criterion 7 Indicator assessment and reporting for the 2010 US National Report on Sustainable Forests was seen as an opportunity to bridge between the original and revised Indicators. To achieve this, we developed a new theoretical approach to describe the status and changes in the SFM Indicators under Criterion 7.

In the following sections, we present the approach developed in the U.S. to analyze the original Criterion 7 Indicators and discuss some of the key findings as well as implications for the next assessment of forest sustainability in the U.S. through the Montreal Process.

4. Indicator analytical methods

4.1 Theoretical model

An understanding of the effectiveness of the legal, institutional, and economic framework for forest conservation and sustainable management first requires knowledge of related policy. Policy may be considered a purposive course of action or inaction that an actor or set of actors takes to deal with a problem (Anderson 2010, Hiedenheimer et al. 1983). Policy statements are the formal written outputs of government or private decisions that express the means for implementing policy goals. Laws and regulations are generally the first formal step to policy implementation, which may also include informational, educational, fiscal, market-based and voluntary mechanisms and applications.

In order to understand and analyze the effectiveness of the legal, institutional, and economic framework for forests in the U.S., we drew from theory and research on policy instruments and their analysis (Sterner 2003, Cubbage et al. 2007), "smart regulation" (Gunningham et al. 1998), forest regulatory "rigor" (Cashore and McDermott 2004), and nonstate governance of sustainable forestry (Cashore et al. 2004). Rooted in this literature, McGinley (2008) developed a theoretical model for analyzing the forest policy *structure* and *approach* of government regulation and non-government forest certification in prospective study countries in Latin America. Policy *structure* refers to the level of obligation on the part of individuals and organizations, or government compulsion (voluntary, mandatory) and the policy *approach* refers to the type of policy or practice employed (prescriptive, process-based, performance-based). This model was developed to examine forest policy directives intended for the forest management unit level. Thus, it was modified for use in our analysis of Criterion 7 Indicators for the U.S.

Implementation of the U.S. Legal, Institutional, and Economic Criterion and Indicators
for the 2010 Montreal Process for Sustainable Forest Management

55

For Criterion 7, the scale of the institutional responses to forest conservation and sustainable management is particularly relevant, since there is wide variation among the 50 U.S. states, not to mention the innumerable local government jurisdictions. Furthermore, many of our U.S. policies and institutions are actually determined by private markets, not government, so this must be considered as part of the analysis of the Criterion 7 Indicators. Therefore, modifications to McGinley's (2008) model included the expansion of policy structure to account for higher level policy mechanisms (non-discretionary/command-and-control; informational/educational; discretionary/voluntary; fiscal/economic; market-based), and adding an approach component for the role of private enterprise in setting institutional policy (Figure 1).

The model displayed in Figure 1 illustrates the range and variation in forest policy mechanisms, approaches, and scales, as characterized by Gunningham et al. (1998); Cashore and McDermott (2004); Cashore et al. (2004); Sterner (2003), and Cubbage et al. (2007). Note that the schema summarized in Figure 1 varies by policy mechanism (often referred to as policy instruments) from command-and-control to market-based, and by approach from prescriptive to private enterprise. To some extent these are continuous scales, not categorical, but we used the categories to make classification and discussion clearer.

We operationalized the theoretical concepts presented in Figure 1 into a "Forest Policy and Governance Matrix" by converting the model into a two-sided classification schema, which we used to classify U.S. SFM laws, institutions, and economic programs under Criterion 7 (Table 2), and to provide comparisons and a meaningful basis for the discussion of each Indicator. This classification schema also fits nicely within the more detailed schema of policy instruments for multi-functional forestry developed by Cubbage et al. (2007), which is presented in Appendix A.

I. Scale			
National	Regional	State	Local
II. Mechanism				
(A) Non-Discretionary/ Command and Control	(B) Informational/ Educational	(C) Voluntary/ Discretionary	(D) Fiscal/ Economic	(E) Market Based
III. Approach			
Prescriptive	Process or Systems Based	Performance or Outcome Based	Private Enterprise

Fig. 1. Forest Policy and Governance Matrix by Geographic Scale, Mechanism, and Approach for the United States

In its application, we added specificity to the Matrix by detailing the types of policy instruments that may be employed through the legal, institutional, and economic framework for forest conservation and sustainable management. These include government ownership, Best Management Practices, payments for environmental services, and forest certification, among many others. The typology of specific policy instruments that we reviewed is listed at the bottom of Table 2 and described in detail in the next section.

Mechanism	Scale: National (N), Regional (R), State (S), Local (L)	Approach			
		Prescrip-tive	Process or Systems Based	Perfor-mance or Outcome Based	Private Enter-prise
Non-Discretionary/ Mandatory[a]					
Informational/ Educational[b]					
Discretionary/ Voluntary[c]					
Fiscal/Economic[d]					
Market Based[e]					

Policy Instruments Possible that Could Be Entered in Each Row of the Table 2 Above:
[a] Laws (L), Regulations or Rules (R), International Agreements (I), Government Ownership or Production (G)
[b] Education (E), Technical Assistance (T), Research (R), Protection (P), Analysis and Planning (A)
[c] Best Management Practices (B), Self-regulation (S)
[d] Incentives (I), Subsidies (S), Taxes (T), Payments for Environmental Service (P)
[e] Free enterprise, private market allocation of forest resources (M), or market based instruments and payments, including forest certification (C) wetland banks (W), cap-and-trade (T), conservation easement or transfer of development rights (E)

Table 2. U.S. Forest Policy and Governance Matrix by Geographic Scale, Mechanism, and Approach

The Forest Policy and Governance Matrix developed for the U.S. corresponds well with the general qualitative indicators developed by the Ministerial Conference on the Protection of Forests in Europe (MCPFE 2003). That Process also categorized forest policy instruments into three similar classes: legal/regulatory, financial/economic, or informational. In addition, the MCPFE schema identifies the main policy area, objectives, and relevant institutions. We include most of these factors in our matrix in similar categories, which we termed policy *mechanisms*.

4.2 Using the Matrix model
In our Matrix (Table 2), *approaches* to forest policy and governance include prescriptive, process- or systems based, performance or outcome based, and private enterprise. A *prescriptive* policy identifies a preventive action or prescribes an approved technology to be used in a specific situation. It generally requires little interpretation on part of the duty holder, offers administrative simplicity and ease of enforcement, and is most appropriate for problems where effective solutions are known and where alternative courses of action are undesirable. However, a prescriptive policy may also inhibit innovation or discourage

Implementation of the U.S. Legal, Institutional, and Economic Criterion and Indicators
for the 2010 Montreal Process for Sustainable Forest Management

57

adaptive management (Gunningham et al. 1998). An example of non-discretionary prescriptive standard is: "Cutting intensity does not exceed 60% of the number of trees per species with a diameter at breast height greater than or equal to 60 cm."

A *process-based* policy identifies a particular process or series of steps to be followed in pursuit of a management goal, such as conservation of endangered species habitat or public involvement in National Forest management planning. It typically promotes a more proactive, holistic approach than prescriptive-based policies. Challenges associated with process-based policies include complicated oversight, compliance 'on-paper' rather than on the ground, and an over-reliance on management systems (Gunningham et al. 1998). An example of discretionary process-based is: "Measures should exist to control hunting, capture and collection of plant and animal species." The fact that there is a process developed also has an embedded assumption that a good process leads to good outcomes, which is often but not always the case.

Performance-based policy specifies the management outcome or level of performance that must be met, but does not prescribe the measures for attainment. It allows the duty holder to determine the means to comply, permits innovation, and accommodates changes in technology or organization. Performance-based policies neither specifically promote nor preclude continuous improvement, and enforcement may require intensive monitoring, analysis, and related resources (Gunningham et al. 1998). An example of non-discretionary performance-based policy is: "The rate of forest products harvested does not exceed the rate of resource growth."

Private enterprise relies on voluntary market exchange to allocate many of the forest resources in the world, both in private markets and for allocation of goods and services on public lands. Many new market-based conservation incentives are being developed as well (Cubbage et al. 2007). Market mechanisms represent both a broad philosophical policy approach — letting the private sector develop policies — and a number of mechanisms or instruments, often supported by government. Markets provide flexibility in individual and firm responses and promote innovation, but outcomes are not directly measured or guaranteed. Furthermore, markets do not ensure or even yield equitable outcomes. In many cases in the U.S. and elsewhere, markets for private goods are deemed best to achieve SFM. In addition, many public policy mechanisms, such as the regulation of no net loss of wetlands or payments for permanent easements to protect forest lands, have involved public-private partnerships to achieve SFM.

In addition to the various *approaches* to policy implementation, there are various *mechanisms or policy instruments* that have been employed to protect and sustainably manage forests. These range from mandatory command-and-control regulations or government ownership to reliance on market-based certification or cap-and-trade to allocate forest resources. Intermediate steps between these approaches include information and education, voluntary, and fiscal or incentive mechanisms. Cubbage et al. (2007) outline these approaches in detail (Table 3), and we relied on that schema to identify specific policy *mechanisms* relevant to each SFM Criterion 7 Indicator.

In using the Forest Policy and Governance Matrix displayed in Table 2, the first column identifies the *mechanism* or instrument through which policies and programs are implemented. The second column denotes the *scale* at which policy is developed and applied. The final four columns show the policy *approach* (prescriptive, process-based, performance-based, private enterprise). Specific policy *instruments* are listed in further detail at the bottom of the table. These are used to add further detail to the *approach* columns, with the most prescriptive policies appearing in the upper left of the matrix and the most voluntary appearing in the lower right.

In the matrix, non-discretionary approaches and instruments would include, laws (L), regulations and rules (R), international agreements (I), and government ownership (G). Informational or educational approaches include education (E), technical assistance (T), research (R), protection (P), and analysis and planning (A). Voluntary approaches include best management practices (B), or self-regulation (S), such as forest certification. Fiscal and economic approaches include incentives (I), subsidies (S), taxes (T), or payments for environmental services (P). Last, free market mechanisms include private markets (P), market based systems such as forest certification (C), wetland banks (W), cap-and-trade (T), and conservation easements (E).

The Criterion 7 analysis for the 2010 US National Report on Sustainable Forests (USDA Forest Service 2011) was seen as an opportunity to bridge between past, current, and future assessments of forest laws, institutions, and policies. The Forest Policy and Governance Matrix that we developed for the 2010 National Report can be utilized, along with the in-depth analysis of previous reporting, to track changes in the status of the Criterion 7 Indicators in future assessments.

For the 2003 National Report on Sustainable Forests, Ellefson et al. (2003) performed detailed analyses and summaries of most Criterion 7 Indicators (USDA Forest Service 2004). We utilized these analyses as the basis for the 2010 Criterion 7 update, examining them through the lens of the Forest Policy and Governance Matrix, and identifying and analyzing any changes in the associated legal, institutional, or economic framework. These combined analyses served to generate the 2010 C7 Indicator reports. The Matrix can be used in future assessments to analyze revisions in Criterion 7, and to assess trends in a systematic manner. This approach also provides a framework for comparing U.S. and other Montreal Process countries at a given point in time.

We used the Forest Policy and Governance Matrix to classify the U.S. legal, policy, and economic approaches to forest conservation and management as described by the Indicators under Criterion 7 of the Montreal Process. We first prepared an initial draft characterizing the U.S. approach to each Indicator according to the relevant variables and cells in the matrix. These draft analyses were reviewed by experts in a set of three public workshops on the U.S. SFM C&I, and well as through an extensive open public comment process.

Based on the political science theory, the draft Forest Policy and Governance matrix, and the public meetings and written reviews, we revised the approach slightly, and the application to various indicators moderately. Then we re-analyzed and applied the matrix to each of the 20 legal, institutional, and economic indicators used for the 2010 report.

To illustrate the application of the Forest Policy and Governance Matrix, Appendix B shows the relevant matrix and associated text published in the U.S. National Report on Sustainable Forests 2010 for Indicator 7.1.d - Extent to which the legal framework (laws, regulations, guidelines) supports the conservation and sustainable management of forests, including the extent to which it encourages best practice codes for forest management. A similar set of matrices and text was published for each of the 20 C7 Indicators in the National Report (Moffat et al. 2011). In using the Matrix, note that each Indicator in Criterion 7, and in the National Report, with a couple of exceptions, had a standard two-page write-up. The Criterion 7 template for each Indicator included a description of what the indicator is and why it is important; the Policy and Governance Matrix with the Relevant Approach,

Implementation of the U.S. Legal, Institutional, and Economic Criterion and Indicators
for the 2010 Montreal Process for Sustainable Forest Management

59

Mechanism and Scale cells completed; a statement of what the indicator shows; and what has changed since 2003.

5. Discussion

The summaries from the 2003 National Report and the Forest Policy and Governance Matrix were used as a framework to discuss each Indicator in Criterion 7 and to make more general observations about the U.S. legal and institutional approach to SFM in the 2010 National Report on Sustainable Forests. Conclusions from this theory-based analysis verify that there is a wide variety of legal, institutional, and economic approaches that encourage sustainable forest management in the United States, at all levels of government. Public laws govern public lands, which comprise about one-third of the nation's forests. They dictate management and public involvement through various detailed approaches and mechanisms. Federal and state laws also provide for technical and financial assistance, research, education and planning on private forest lands, but do not prescribe specific actions or standards. However, at the state and local level, in many cases, laws do prescribe specific management actions or standards, such as state forest practice acts, prescribed burning laws, water quality standards, and local zoning regulations.

Federal and state environmental laws protect wildlife and endangered species in forests on all public and private lands. They regulate or promote (best) forest practices to protect water quality, air quality, or other public goods, varying significantly by state. Private markets allocate forest resources on most private forest lands, and even governments use markets for making timber sales, leasing lands for minerals, contracting with private concessionaires for tree planting, or providing recreation services. Many new market based mechanisms, including forest certification, wetland banks, payments for environmental services, conservation easements, and environmental incentives are also being developed to implement sustainable forest management and conservation on private and public lands in the United States.

The effectiveness of the Criteria and Indicators in achieving SFM does rely ultimately on value-based politics, which determine the effectiveness of policies and institutions. The Matrix can enhance the rigor and clarity of this discussion and analysis, help clarify gaps and weaknesses in our institutions, and identify opportunities for improvement in the pursuit of sustainable forest management. Note that the Matrix and associated discussion are intended to summarize the institutional context, not to make policy recommendations. Other parts of the National Report and related subsequent implementation efforts such as that by the Pinchot Institute (Sample et al. 2006) can provide appropriate means of identifying policy responses.

The 2009 Montreal Process modifications to Criterion 7 and its Indicators are expected to better facilitate assessments of the current status and trends in forest laws, institutions, and policies. The revised 10 C7 Indicators to be used in the next round of reporting and beyond stem from the original 20 Indicators, but are more succinct and objective. While they are still more apt to be described qualitatively than measured quantitatively, they are expected to improve measurement and reporting.

Based on the revised 2011 Criterion 7 Indicators, future analysts will be able to summarize existing laws and polices supporting SFM; effects of taxation or incentives; the relative strength of tenure rights; programs and cooperative efforts; public participation; and monitoring and reporting. The Policy and Governance Matrix developed for the 2010 US

National Report on Sustainable Forests can be used to categorize these efforts and subsequent data summaries and legal or policy analyses can add depth to the theoretical framework.

6. Applications

The usefulness of the original 2010 Criterion 7 Indicators and the Forest Policy and Governance Matrix rests on their abilities to condense and convey national, regional, and state information about the policies, laws, and institutions promoting the conservation and sustainable management of U.S. forests. Like the other C&I, Criterion 7 and its Indicators represent an attempt to track the status and trends of forest sustainability for the nation. However, as documented here, the social and legal bases for sustainability are difficult to quantify. We tried to at least make the analysis of this Criterion and its Indicators more consistent and objective through a theoretically-based approach.

Many of the Montreal Process C&I are being used beyond the mere reporting of status and trends, and indeed are leading to program or policy changes and development. Examples include the identification of forest health problems or tracking of fire occurrences and conditions, which then lead to new programmatic responses. The C&I reports for several countries also form the basis for national program development and monitoring, such as for implementation of programs to achieve Reduced Emissions from Degradation and Deforestation (REDD). Comprehensive C&I assessments provide the data and structural platform to design and implement national REDD programs, and in some cases even the structure for forest management level measurement and monitoring.

Description, monitoring, and tracking of the C7 Indicators can also assist in identifying and improving national or state programs for SFM. For example, bilateral trade agreements often require demonstration of sustainable forest practices, which can be evidenced by laws, institutions and policies tracked in Criterion 7, by the U.S. and by our Montreal Process trading partners. Questions about environmental laws and illegal logging addressed in Criterion 7 have become key issues in trade of forest products. These Indicators also are relevant for cross-country comparisons. As the 10 new simplified Criterion Indicators are implemented, the comparison within and among countries will become even more useful. Similarly, so will our Forest Policy and Governance Matrix, or some adaptation of that conceptual framework.

In general, the characterization/categorization of legal and institutional aspects related to SFM as required by Criterion 7 is not a measure of their adequacy for forest conservation and management. Though this same tact (i.e., 'just the data') is taken for the Indicators associated with Criteria 1 through 6, for many of those Indicators the linkage between the data and sustainability can be surmised or, at least, considered. This link is more difficult to make with characterizations of forest policies, laws, and institutions. Perhaps the best use of the C7 analysis is a more explicit and comprehensive categorization of the legal and institutional framework for forests that leads to a better understanding of related policy, law, and institutions, and thereby provides a more complete and transparent basis for assessing the overall framework in regards to actual outcomes and, ultimately, to forest sustainability.

The Criterion 7 indicators do not measure sustainability directly, but address the social components of sustainable development. To some extent, they are the tools used to achieve sustainable forest management. The ecological and even social SFM C&I help directly

Implementation of the U.S. Legal, Institutional, and Economic Criterion and Indicators
for the 2010 Montreal Process for Sustainable Forest Management

61

measure and monitor the status of SFM. Thus in the Montreal Process C&I construct, Criteria 1 to 6 are mostly objective measures of forest sustainability, and Criterion 7 is the assessment of the institutions that help achieve sustainability. The implementation and effectiveness of these laws and institutions will determine how well sustainable forest management is achieved.

Consistently using an analytical tool like the Forest Policy and Governance Matrix in future assessments would facilitate measurements of changes in policy over time, as well as cross-country comparisons, and would potentially permit assessments of related results. The key will be in detecting variance both in terms of matrix coding and in terms of forest impacts and outcomes. These sorts of comparisons (i.e., over time, cross-country) would permit a more substantive characterization of forest policy approaches, to determine, for example, if the U.S. relies more/less on economic incentives to promote SFM than in the past, or more/less than other countries, and given links to other forest measures, may permit associations with changes in forest conservation and management.

7. Conclusion

In the 2010 U.S. National Report on Sustainable Forests, we developed a theory-driven classification scheme to discuss each of the Indicators of SFM in Criterion 7. This approach relied on existing available data and information that was examined through the lens of the Forest Policy and Governance Matrix to measure and monitor legal, institutional, and policy trends related to SFM in the U.S.. The effectiveness of these C&I in achieving SFM does rely ultimately on normative measures about the effectiveness of policies and institutions. Moreover, there is significant debate regarding which forest policies are "best" for achieving SFM, particularly in different countries and biophysical and social contexts. Our analytical approach can enhance the rigor and clarity of this discussion and analysis, help clarify gaps and weaknesses in the legal and institutional framework, and identify opportunities for improvement to achieve SFM.

It is important to note that the intent of Criterion 7 is to provide an objective measurement of the status of laws, policies, and institutions that support forest conservation and management in each country, and perhaps allow comparisons among countries. This is nominally a "positive" or value-free analysis, not a normative assessment designed to make policy recommendations. This is a subtle distinction, since each Indicator reflects specific elements of the value-laden policies that governments choose to enact. Criterion 7 and its Indicators are meant to reveal the status of public policies related to forest conservation and management. Decisions on the adequacy of these public policies in promoting SFM are left to high-level government policy-makers and the relevant legislatures and related interest groups.

In most countries, agency personnel are charged with implementing legislative, executive, and judicial policy decisions, not advocating for changes, even through analytical assessments like those derived from SFM C&I applications. This requires that the C&I be analyzed and reported judiciously in each country report. In fact, the U.S. report primarily focused on the technical findings of the seven Criteria and 64 Indicators, such as forest area trends, forest health issues, carbon storage, forest fragmentation, timber and nontimber market values. And, though the report identifies the "implications of the findings for policy and action", it purposefully does not make policy judgments or recommendations. Nonetheless, an assessment of the status and change in forest policy, law, and institutions through the Criterion 7 Indicators provides information to decision- and policy-makers, who are then authorized to determine if the legal and institutional framework at various levels is adequately addressing forest conservation and sustainability, or if changes should be made, and whether that can be afforded in the current and probably enduring times of budget austerity.

Overall, this new approach to analyzing the 2010 and perhaps future Criterion 7 Indicators provides a better understanding over time of the ways in which policy, legal, and institutional capacity affects forest sustainability. The outcome of this process will determine the extent to which the work on Criterion 7 presented in this document becomes a foundation for future reporting. In any case, the analysis presented here provides a consistent and useful way of characterizing and understanding a broad and complex topic area.

Appendix A. Selected Policy Instruments for Multi-Functional Forestry (Cubbage et al. 2007)

Government Ownership and Planning	Government Regulation	Subsidies & Protection	Education & Research	Private Markets	Private/ Public Project Financing	Private/Public Market Development
Land ownership	Best practices	Plantations	*Education*	*Land Ownership/ Management*	*Financing and grants*	Tradable development rights
National	Harvesting, roads	Timber stand improvement	Professional	Small private	International bank Loans	Conservation easements
Community	Illegal logging	Income tax reduction	Continuing	Industrial	Debt-for-nature swaps	Concession/ extraction quotas
Native/indigenous	Water quality and quantity	Property tax reduction	Public	Timber investment organizations	Venture capital funds	Tradable protection rights
Production	Wildlife, biodiversity	Forest industry & manufacturing	Landowner	Environmental organizations	National forestry funds	Water resource use charges
Timber products	Endangered species	Ecosystem management	Logger and worker	Cooperatives	Policy/ business guarantees	Bioprospecting fees
Nontimber products	Landscape effects	Environmental services	*Research*	*Goods and Services*	Conservation trust funds	Payments for environmental services
Final products	Aesthetics	Fire protection	Federal	Products	Environmental protection funds	Payments for environmental degradation
Services &Amenities	Conversion	Insect & disease protection	State	Services	Securitization	Carbon offset payments
Recreation	Workers/safety/ pay	Invasive species	Forestry schools	Amenities	Grants by philanthropies, NGOs	Clean Development Mechanism
Environmental Services	Community benefits/impacts	Trespass, theft, illegal logging	Other academic disciplines	*Financing*	*Joint management arrangements*	
International Fora and SFM Processes	International trade agreements	Forest law enforcement & governance	Private industry	Banks/loans/ credit	Contracting, leasing, joint	
SFM Criteria & Indicators			Non-government organizations	Foreign direct investment	Build Operate Transfer	
UN Forum on Forests				Forest certification	Build Own Operate	

Implementation of the U.S. Legal, Institutional, and Economic Criterion and Indicators
for the 2010 Montreal Process for Sustainable Forest Management

63

Appendix B. Verbatim Text of Indicator 7.48 from The National Report on Sustainable Forests, 2010

Indicator 7.48 - Extent to which the legal framework (laws, regulations, guidelines) supports the conservation and sustainable management of forests, including the extent to which it encourages best practice codes for forest management

What is the indicator and why is it important?

Forest management practices that are well designed are fundamental to the sustainability of forest resources. At all levels (stand, landscape, local, regional, national, global), forests depend on the application of forest practices that are capable of ensuring sustained use, management, and protection of important social, economic, and biological values. Well-founded best practice codes, and the forest management practices that comprise them, can ensure sustained forest productivity for market goods; protection of ecological values; and protection of the various social, cultural, and spiritual values offered by forests. They can be among the most important tools for responding to national trends and conditions involving forests.

Policy and Governance Classification

Mechanism	Scale: National, Regional, State, Local	Approach			
		Prescriptive	Process or Systems Based	Performance or Outcome Based	Private Enterprise
Non-Discretionary/ Mandatory[a]	N,S,L	L,R,G	L,R,G	L,R	
Informational/Educational[b]	N,S,L	P,T,R	E,T,R	E,T,R	
Discretionary/Voluntary[c]	N,S	B	B	B	B,S
Fiscal/Economic[d]					
Market Based[e]	N,S,L				C

[a]Laws (L), Regulations or Rules (R), International Agreements (I), Government Ownership or Production (G)
[b] Education (E), Technical Assistance (T), Research (R), Protection (P), Analysis and Planning (A)
[c] Best Management Practices (B), Self-regulation (S)
[d] Incentives (I), Subsidies (S), Taxes (T), Payments for Environmental Service (P)
[e] Free enterprise, private market allocation of forest resources (M), or market based instruments and payments, including forest certification (C) wetland banks (W), cap-and-trade (T), conservation easement or transfer of development rights (E)

What does the indicator show?

National, state, and local government landowners, as well as all private landowners, have various levels of recommended or required forest best management practices (BMPs). BMPs may be implemented through educational, voluntary guidelines, technical assistance, tax incentives, fiscal incentives, or regulatory approaches.

Ellefson et al. (2005) provide detailed summary of BMPs, albeit for 1992, but it can provide a guide for types of programs now. More than 25 states have regulatory forestry BMPs to protect water quality and to protect landowners from wildfire, insects, and diseases. Almost all states (≥ 45) have educational and technical assistance programs for BMPs about water quality, timber harvesting methods, protecting wildlife and endangered species; and more than 40 have such programs to enhance recreation and aesthetic qualities.

Even states that do not have legally required BMPs often have water quality laws intended to control surface erosion into water bodies of the state, and can be used to enforce BMP compliance. Local governments also implement BMPs for private forest lands, along with other land use controls on development, agriculture, or mining.

BMPs may be prescriptive and mandatory, as required in the state forest practice laws of all the states on the West Coast and many in the Northeast; may require that forest managers and loggers follow specific processes, such as in Virginia; or may be performance or outcome based, ensuring that water quality is protected, such as in North Carolina.

BMPs may cover a variety of practices, such as timber harvest, road construction, fire, site preparation and planting, and insect and disease protection. They also may cover diverse natural resources to be protected, such as water quality, air quality, wildlife, endangered species, or visual impacts.

While BMPs are pervasive, differences of opinion exist about their effectiveness. Almost all forestry compliance surveys have found a high overall rate of compliance for most landowners, but environmental groups contend that many individual practices, such as road-building or wildlife habitat impacts, remain problematical.

The federal government and most states provide detailed technical assistance for information and education about BMPs, as well as research about efficacy, benefits, and costs. The private sector including forest industry, large timberland investors, nonindustrial private forest owners, and forest consultants have been actively involved in development and promotion of BMPs. BMP compliance also is required as part of the standards of all three major forest certification standards in the U.S.—the Sustainable Forestry Initiative, Forest Stewardship Council, and American Tree Farm System.

What has changed since 2003?

Voluntary and regulatory state best management practices for forestry have continued to evolve and improve since 2003. They have been evaluated periodically through on-the-ground effectiveness surveys, and periodically revised. Their scope has been extended in some states to cover more than just timber harvesting and roads to include wildlife, landscape level effects, or aesthetics. Enforcement has increased through inspections, even in states with voluntary BMPs. Several states also have issued separate BMPs for biomass fuel harvesting. And BMPs are now explicitly required under all forest certification systems in the United States.

8. References

Anderson, J. (2010). *Public Policy Making: An Introduction, 7th ed.* Wadsworth, ISBN: 978-0-618-97472-6, Boston, USA.

Cashore, B., Auld, G. and Newsom, D. (2004). *Governing Through Markets: Forest Certification and the Emergence of Non-State Authority*. Yale University Press, ISBN 0-300-10109-0, New Haven, USA.

Implementation of the U.S. Legal, Institutional, and Economic Criterion and Indicators
for the 2010 Montreal Process for Sustainable Forest Management

65

Cashore, B. and McDermott, C.L. (2004). Global Environmental Forest Policy: Canada as a constant case comparison of select forest practice regulations. International Forest Resources. Victoria, B.C.

Cubbage, F., Harou, P. and Sills, E. (2007). Policy instruments to enhance multi-functional forest management. *Forest Policy and Economics* 9:833-851.

Ellefson, P.V., Hibbard, C.M., Kilgore, M.A., and Granskog, J.E. (2005). Legal, Institutional, and Economic Indicators of Forest Conservation and Sustainable Forest Management: Review of Information Available for the United States. Gen. Tech. Rep. SRS-82. Asheville, NC: U.S. Department of Agriculture, Forest Service, Southern Research Station.

Gunningham, N., Grabosky, P., and Sinclair, D. (1998). *Smart Regulation: Designing Environmental Policy*. Clarendon Press, ISBN: 0198268572, New York.

Humphreys, D. *Logjam: Deforestation and the Crisis of Global Governance*. Earthscan. ISBN 978-1-84407-611-6, London.

Hiedenheimer, A., Heclo, H., and Adams, C. 1983. *Comparative Public Policy: The Politics of Social Choice in Europe and America*. St. Martin's, ISBN 031215366X, New York.

McDermott, C.L., Cashore, B and Kanowski, P. (2010). *Global Environmental Forest Policies: An International Comparison*. Earthscan, ISBN 978-1-84407-590-4, London.

McGinley, K.A. (2008). Policies for Sustainable Forest Management in the tropics: governmental and non-governmental policy outputs, execution, and uptake in Costa Rica, Guatemala, and Nicaragua. Ph.D. dissertation. North Carolina State University. September 2008.

MCPFE. (2003). Improved Pan-European Indicators for Sustainable Forest Management. Adopted by the MCPFE Expert Level Meeting 7-8 October 2002, Vienna, Austria. Ministerial Conference on the Protection of Forests in Europe. MCPFE Liaison Unit Vienna. 6 p.

Moffat, S., Cubbage, F. and McGinley, K. (2011). Legal, institutional, and economic framework for forest conservation and sustainable forest management. p. II-107-II-134. In: National Report on Sustainable Forests—2010. United States Department of Agriculture Forest Service Publication FS-979. June 2011. Accessed at: http://www.fs.fed.us/research/sustain/2010SustainabilityReport/documents/20 10_SustainabilityReport.pdf. June 30, 2011.

Montreal Process. (2009). *Criteria and indicators for the conservation and sustainable management of temperate and boreal forests*. Fourth Edition, October 2009. 48 p.

Sample, V. A., Kavanaugh, S.L., and Snieckus, M.M. eds. 2006. Advancing Sustainable Forest Management in the United States. Pinchot Institute for Conservation. Washington, D.C. Accessed at: http://www.pinchot.org/pubs/. 29 June 2009.

Sterner, T. 2003. *Policy Instruments for Environmental and Natural Resource Management. Resources for the Future*, ISBN 1-891-853-13-9, Washington, D.C., USA.

USDA Forest Service. 2004. National Report on Sustainable Forests—2003. United States Department of Agriculture, Forest Service. FS-766. Washington, D.C.

USDA Forest Service. 2011. National Report on Sustainable Forests—2010. United States Department of Agriculture, Forest Service. FS-979. Washington, D.C.

Wijewardana, D. 2008. Criteria and indicators for sustainable forest management: The road travelled and the way ahead. *Ecological Indicators* 8(2008):115-122.

Section 2

Decision Making Tools

5

How Timber Harvesting and Biodiversity Are Managed in Uneven-Aged Forests: A Cluster-Sample Econometric Approach

Max Bruciamacchie, Serge Garcia and Anne Stenger
Laboratoire d'Economie Forestière, INRA/AgroParisTech-ENGREF
France

1. Introduction

Nonindustrial private forest (NIPF) landowners have been shown to be more multi-objective by nature than industrial landowners: they give more importance to standing timber and forestland for the amenity values they provide (Newman & Wear, 1993). Among analyses of forest landowner behaviour, the household production framework recognises the benefits associated with forest amenities, as first applied by Binkley (1981). These non-market services are jointly produced with timber and are a determinant in the landowner's utility function.

NIPF landowners comprise close to 70% of land ownership in many U.S. states and significant land holdings throughout Europe (Amacher et al., 2003). In France, almost 75% of the total forestland is privately owned, and 96% of private landowners are nonindustrial. In this article, we investigate the joint production of timber and biodiversity for NIPF landowners using a micro-econometric household production model.

Even though our model is situated within a standard framework where a non-marketed good is jointly produced with timber products, we consider here that biodiversity is not totally disconnected from market strategies. Biodiversity is measured by the diversity of tree species. This assumption is based on the theory of coevolution introduced by Ehrlich & Raven (1964). Coevolution acts as an evolutionary engine and a vehicle for biological diversification. Thus, the diversity of trees or plants may not only tend to increase the diversity of insects and animals, but the converse may also be true. In our model, tree diversity is a determinant of consumer satisfaction and a joint product in the profit-maximisation problem. Tree diversity has an additional impact: it is closely related to some market aspects since the different species have different monetary values. The forest landowner can decide to favour one tree species over another, depending on its value on the market. Conversely, he can make the choice of species diversification to cope with the volatility of timber prices.

We focus on a complete set of forest landowners' decisions in uneven-aged forests where landowners are assumed to value the tree diversity of their forests, as well as timber harvesting. Our economic model is based on the maximisation of their utility that depends on the revenues from harvesting and tree diversity with respect to technological and budgetary constraints. The global objective of the paper is to explain the links between some of the harvest strategies of forest owners, unit price variability and the observed diversity of trees.

More precisely, we analyse: (1) their demand for species diversity and their timber supply; and (2) the joint production of timber and species diversity. Timber supply and amenity demand functions are derived using first-order conditions of the maximisation problem for the landowner.

The behaviour of the forest owner is also strongly dependent on the characteristics of the forest blocks in question. Moreover, his/her harvesting strategy should differ according to the tree species and its value (depending itself on the quality and the diameter of the trees). The issue of heterogeneity in this case is crucial and its omission may result in consequent biases in the estimation stage. The estimation of timber supply and diversity demand is made using a database on uneven-aged forests in France for which several economic and ecological variables are regularly collected. This database typically concerns several forest blocks within which different tree species cohabit. This makes it possible to consider the forest owner within a multi-product framework where each product corresponds to a particular tree species.

2. Methods

2.1 Biodiversity and the economic model

In the literature on NIPF landowners, recent models of timber supply have included non-monetary returns or amenities (Binkley, 1981; Hyberg & Holthausen, 1989; Max & Lehman, 1988; Pattanayak et al., 2003; 2002). The idea is to better understand the trade-off between timber harvesting and amenity benefits.

In this study, we attempt to understand forest owners' decisions concerning timber harvesting and biodiversity. Indeed, different tree species have different monetary values, and the forest landowner has several alternatives: to favour one tree species over another, depending on its market value, or to diversify the tree species in order to cope with the volatility of timber prices.

Our definition of biological diversity may appear to be restrictive due to the sole inclusion of trees (instead of global biodiversity). Nevertheless, tree diversity accounts for a large part of biodiversity: it is generally accepted that the mixture of species is the guarantee of a certain degree of diversity of other living communities (for invertebrates, see Greatorex-Davies et al. (1993), and for bats, see Mayle (1990)). This is the principle of coevolution (Ehrlich & Raven, 1964). The diversity of trees or plants may not only tend to increase the diversity of insects and animals, but the converse may also be true.[1] Even if the extrapolation of tree diversity to global biological diversity is still in debate, this makes it possible to take both biodiversity and strategies on the timber market into account with only one indicator. Furthermore, there is no consensus about the choice of the diversity indicator. This is why several measures were tested in our model.

We, in fact, used two notions, richness and diversity, the latter being the Shannon diversity index computed as $H = -\sum_h p_h \ln p_h$, where h represent a species. Three diversity indices were calculated:

1. Tree richness, designated by $RICH$, is computed as the number of species in the forest compartment. This is the simplest and the most intuitive index used to measure biodiversity. However, this measure strongly depends on the area surveyed.

---

[1] Many references exist on this topic, see Lähde et al. (1999), Barbier et al. (2008), Schuldt et al. (2008), McDermott & Wood (2009), among others.

2. The Shannon diversity index on the basis of number, designated by $SHANN$, is computed from the number of stems (n_h) with $p_h = \frac{n_h}{\sum_h n_h}$.

3. The Shannon diversity index on the basis of volume, referred to as $SHANV$, is expressed in volume v_h: $p_h = \frac{v_h}{\sum_h v_h}$. The Shannon diversity index based on number is often used by ecologists, but the Shannon diversity index based on volume is more effect for characterising the crown size of different species.

In our model, tree diversity is a determinant of consumer satisfaction and a joint product in the profit-maximisation problem. The landowner i is represented in the framework of the household production function by a utility function that depends on the total income and non-pecuniary attributes:

$$U_i = U(I_i, z_i),\qquad(1)$$

where I_i represents the total income of the landowner i and z_i is the forest biodiversity. The forest landowner faces a budget constraint where the total income is the sum of timber production profit π and exogenous income E:

$$I_i = \pi_{ij} + E_i.\qquad(2)$$

The timber profit π_{ij} depends on timber production y_{ij} sold at the price p_{ij}, where the subscript j designates the tree species. The profit function is the difference between the timber revenue and the multi-product cost function related to the production of the (marketable) timber output y_{ij} and the tree diversity z_i conditional on some exogenous variables x_{ij} (including forest capital and ecological variables). It can be written as:

$$\pi_{ij} = p_{ij} \times y_{ij} - C(y_{ij}, z_i, x_{ij}).\qquad(3)$$

Timber production y_{ij} and tree diversity z_i are linked by the following transformation function:

$$T(y_{ij}, z_i, x_{ij}) = 0.\qquad(4)$$

The forest landowner has to choose the level of decision variables (i.e., y, z and I) that maximizes the utility function (1) subject to constraints (2) and (3). This utility maximisation problem can be solved by substituting these constraints into the utility function. The resolution is done in two steps: the household first selects the optimal level of I and z and then chooses the level of production y. In order to obtain explicit solutions to this problem, we have imposed some simple functional forms on our model. We chose a Cobb-Douglas form for the utility and cost functions. With these particular functional forms and by deriving with respect to y, we obtain the timber supply function that depends on timber price p, non-timber product z and other variables x. Expressing the first-order condition in log-linear form, we find the following timber supply function:

$$\ln y_{ij} = \alpha_0 + \alpha_1 \ln p_{ij} + \alpha_2 \ln z_i + \alpha_3 \ln x_{ij},\qquad(5)$$

where the unknown parameters α are to be estimated. Note that α_1 represents the price elasticity of supply. If α_1 is respectively $<, =$ or > 1 then the supply is price inelastic, unit-elastic or price-elastic. α_2 measures the trade-off between tree harvesting and diversity in terms of elasticity. If α_2 is negative, there is a substitution effect, whereas a positive sign is synonymous with complementarity.

Entering the equation (5) in the utility function and deriving it with respect to z give us the diversity demand. Transforming it into log-linear form, we have:

$$\ln z_i = \beta_0 + \beta_1 \ln p_{ij} + \beta_2 \ln x_{ij}, \tag{6}$$

where β are the unknown parameters of the demand function to be estimated. β_1 represents the elasticity of diversity demand with respect to timber price. If β_1 is respectively $<, =$ or > 1 then the diversity is inelastic relative to the timber price, unit-elastic or price-elastic.

2.2 The econometric approach

A two-step estimation procedure is implemented by first estimating the diversity demand equation (at the forest level), followed by the timber supply equation in which the predicted value of diversity is entered as a regressor.

Harvest observations collected for different tree species in different forests lead to the use of methods specific to cluster sampling (Wooldridge, 2003). However, the diversity of tree species is observed at the forest compartment level and is therefore cluster-invariant. Supposing that all variables are exogenous, the tree diversity demand equation (6) is estimated by the Ordinary Least Squares (OLS) method.

Cluster specificity is taken into account in the estimation of the timber supply equation (5). The units within each cluster (or forest) may be correlated, whereas independence across clusters is assumed. Specific methods applied to Fixed Effect (FE) and Random Effect (RE) models make it possible to control for unobserved forest heterogeneity while studying the effects of factors that vary across species and forests (e.g., price), and others specific to forests (e.g., tree species diversity). Moulton (1986) shows the consequences of inappropriately using OLS estimation in the presence of random group effects. In particular, he demonstrates that the OLS standard errors that are not adjusted in this case are biased.

Consider the following timber supply cluster-sample equation:

$$y_{ij} = \alpha + X_{ij}\beta + Z_i\gamma + u_{ij}, \quad i = 1,\ldots,N, \quad j = 1,\ldots,J_i, \tag{7}$$

where i indexes the "cluster" (or forest), j indexes individual observations within the cluster (or tree species). There is a total number of N clusters. The number of species is not the same throughout the different forests i, so that J (i.e., the number of species in the case of balanced data) is indexed by i. The total number of observations is $n = \sum J_i$. Harvest in the forest i of the tree species j is designated by y_{ij}. X_{ij} is a $(1 \times K)$ vector of explanatory variables that vary with respect to i and j. Z_i contains L explanatory variables that only depend on the cluster i. u_{ij} is the error term. α is the constant, and β and γ are the parameter vectors associated with the X and Z to be estimated, respectively.

We consider the following unbalanced one-way error component:[2]

$$u_{ij} = \mu_i + \epsilon_{ij}, \quad i = 1,\ldots,N, \quad j = 1,\ldots,J_i, \tag{8}$$

---

[2] Only five species are observed and not within all forests. We can therefore not implement a two-way error component regression model. Moreover, each forest is observed only once since we only have cross-section data.

where μ_i is the cluster specific effect, and ϵ_{ij} represents the remaining unobservables. μ_i and ϵ_{ij} are assumed to be independent and respectively i.i.d. $(0, \sigma_\mu^2)$ and $(0, \sigma_\epsilon^2)$. In matrix form, the one-way cluster model can be written as:

$$y = \alpha \iota_n + X\beta + Z\gamma + u$$
$$= R\delta + u,$$

(9)

where $u = R_\mu \mu + \epsilon$, with $R = (\iota_n, X, Z)$ and ι_n a vector of n ones. y and R are of dimensions $n \times 1$ and $n \times (1 + K + L)$. $\delta' = (\alpha', \beta', \gamma')$ is the vector of parameters to be estimated. Finally, $R_\mu = diag(\iota_{J_i})$ with ι_{J_i} is a vector of ones of dimension J_i.

Supposing that all variables are exogenous, the equation can first be estimated by pooled OLS from the unbalanced data. The OLS estimator is trivially given by $\hat{\delta}_{OLS} = (R'R)^{-1}R'y$. It is unbiased and consistent. However, according to the method proposed by Pepper (2002), we use an estimate of the asymptotic variance matrix that is robust to heteroscedasticity and within-cluster correlation of arbitrary forms: $Var(\hat{\delta}_{OLS}) = (R'R)^{-1} \left(\sum_{i=1}^{N} R_i' \hat{u}_i \hat{u}_i' R_i \right) (R'R)^{-1}$, where $\hat{u}_i$ is the $N \times 1$ vector of OLS residuals $(Y_i - \hat{\delta}_{OLS} R_i)$.

Other consistent methods exist (some of which are more efficient), which make it possible to take the presence of unobserved effects in the error term into account. Cluster samples and panel data sets (where i represents individuals and j time periods) can be treated with similar methods (FE and RE models). In our case, the database has the same structure as an unbalanced panel data set. This is why we based our estimation method on the work of Baltagi & Chang (1994).

We can first consider that μ_i represents the unobserved heterogeneity related to the forest, and treat it as a constant parameter to be estimated for each cluster i. If the fixed effects are correlated with the explanatory variables, there is an endogeneity problem that implies a biased estimator of parameters α, β and γ. We can obtain a consistent estimator of β by removing these effects with a suitable transformation (within-group transformation). However, an important drawback is that the parameters (γ) associated with cluster-invariant variables cannot be identified. The within-group transformation matrix for the (unbalanced) cluster-sample case is $Q = diag(E_{J_i})$. $E_{J_i} = I_{J_i} - \frac{\iota_{J_i} \iota_{J_i}'}{J_i}$, where I_{J_i} is an identity matrix of dimension J_i. The Within (or FE) estimator of β is:

$$\hat{\beta}_{FE} = (X'QX)^{-1}X'Qy,$$

(10)

under the assumption of non correlation between ϵ and X. A drawback of this method is that γ cannot be identified because the variables Z disappear after within transformation.

In order to take any possible autocorrelation or heteroscedasticity into account, Arellano (1987) proposes the following variance-matrix estimator:

$$Var(\hat{\beta}_{FE}) = (X'QX)^{-1} \left(\sum_{i=1}^{N} QX_i' e_i e_i' QX_i \right) (X'QX)^{-1},$$

with $e_i = Qy - QX\hat{\beta}_{FE}$, which is fully robust.

If the specific effects are assumed to be non-correlated with the explanatory variables, then a random effects (Generalised Least Squares, GLS) estimation can be used. Even if OLS estimators provide consistent parameters, a heteroscedasticity-consistent variance matrix is necessary. The effect μ_i is now treated as a (cluster-specific) error term and assumed to be i.i.d. $(0, \sigma_\mu^2)$. In this model, we can identify all coefficients related to all variables (including those that are cluster-invariant). Hence, the matrix of explanatory variables is now $R = (\iota_n, X, Z)$. The vector of parameters $\delta' = (\alpha', \beta', \gamma')$ and the variance components $(\sigma_\mu^2, \sigma_\epsilon^2)$ are estimated. The variance-covariance matrix of error terms u is $\Omega \equiv E(uu') = \sigma_\epsilon^2 \Sigma$, where $\Sigma = I_n + \rho Z_\mu Z'_\mu$, with I_n an identity matrix of dimension n and $\rho = \frac{\sigma_\mu^2}{\sigma_\epsilon^2}$. The GLS (or RE) estimator is:

$$\hat{\delta}_{RE} = (R'\Omega^{-1}R)^{-1}(R'\Omega^{-1}y). \tag{11}$$

The variance of the RE estimator is: $Var(\hat{\delta}_{RE}) = \sigma_\epsilon^2 (R'\Omega^{-1}R)^{-1}$. Several methods of estimation of variance components $(\sigma_\mu^2, \sigma_\epsilon^2)$ exist. However, the solution the most often chosen is the method of Swamy & Arora (1972) by using the Within and Between residuals.

The RE estimator is asymptotically more efficient than pooled OLS under the usual RE assumptions. However, if the cluster effects are correlated with μ_i are correlated with X or Z, this estimator is not consistent. This possible endogeneity can be tested for by performing a Hausman test. The Hausman test statistic is: $(\hat{\beta}_{FE} - \hat{\beta}_{RE})'[Var(\hat{\beta}_{FE}) - Var(\hat{\beta}_{RE})]^{-1}(\hat{\beta}_{FE} - \hat{\beta}_{RE})$. Under the null hypothesis, this statistic has an asymptotic chi-square distribution with a number of degrees of freedom equal to the number of cluster-variant variables (K).

3. Results and discussion

3.1 Data sources and characteristics

The database of the AFI network (*Association Futaie Irrégulière* - Uneven-aged forest network) was used. Uneven-aged forest management is characterised by two fundamental principles: the use of natural dynamics of the ecosystem and the individual treatment of each tree. The first principle implies the use of all tree species on the site: forests are always mixed-species (with variations depending on the acidity of the soils). The second principle means that each tree is examined in order to assess its different functions (e.g., value-added wood, aesthetic aspect). Hence, the decision of tree harvesting or conservation does not result from the stand age but rather from its functionality: Does this tree "pay" for its place? (Bruciamacchie & de Turckheim, 2005). Uneven-aged forest management is practised in numerous forests worldwide with a multitude of variations in terms of species composition and stand structures under local ecological, social and economic constraints.

The AFI network consists of 68 compartments in the northern part of France. The compartment is the management unit for uneven-aged forests (whereas the whole forest is the unit considered for even-aged forests) and corresponds to a block that varies from 5 to 15 ha. One compartment is made up of ten permanent plots that make it possible to monitor the individual growth of approximately 200 trees per compartment. These compartments also make it possible to monitor poles, coppice and regeneration. Some of them are good examples of successful transitions between even-aged and uneven-aged stands. Our sample is made up of forests whose stands are well-balanced in terms of forestry (consistent harvesting), which makes it possible to handle economic data that are uniform on the long term.

As mentioned above, we consider a forest owner who maximises his utility that is a function of total income and diversity. The forest owner decides on the main orientations of his/her forest management (e.g., level of revenues, distribution of species, risks concerning species management).   However, we wanted to introduce an important characteristic of forest management into the empirical model: in practice, forests are managed by the "owner/forest manager" pair. Indeed, the owner often delegates the management to a forest manager who implements the owner's choices and can thus have an influence on the harvesting decision and the distribution of species. This is why we include dummies that proxy the identity of the manager (see below).

Among the 68 compartments, 39 were selected because all of the information in all of the categories of variables was available. We classified tree species into five classes: oak, beech, precious broad-leaved trees, other broad-leaved trees and conifers.   These five classes of species are not observed in all of the compartments, so that the total number of observations in our sample is 102.[3]

However, the number of species is greater and we compute the diversity for each compartment from the total number of species (varying from 2 to 14 in our sample).   As presented above, we calculate three diversity indices. The first index used is tree richness, designated by $RICH$, simply computed as the number of species in the forest compartment. The last two are Shannon diversity indices computed as $H = -\sum_h p_h \ln p_h$, where h represent a species.[4] We compute a Shannon diversity index on the basis of number ($SHANN$) and a Shannon diversity index on the basis of volume ($SHANV$), already defined above.

The variables used in the model are the following:

- Variables observed per compartment and broken down by species: harvested volume (y), unit price (p),[5] stock inventory (INV), volume increment ($VOLINCR$).

- Percentage of quality ($QUAL\%$) and average diameter ($DIAM$) are measured for standing timber.

- At the compartment level, seven dummy variables (from $ST1$ to $ST7$)) are built for seven different ecological conditions ranging from the more basic to the more acid soils. In fact, this set of dummies represents an ecological indicator built from the variables, pH and moisture.[6]

- The type of owners is represented by four dummy variables: institution ($DUMO1$), individual ($DUMO2$), group of owners ($DUMO3$) or joint ownership ($DUMO4$).

- The owner often delegates the management to a forest manager. He/she implements the owner's decisions but can have an influence on the distribution of species. Dummies $DUME1$ to $DUME10$ are used for the manager. $DUME10$ is the remaining sum of managers that are in charge of only one forest compartment.

Descriptive statistics are reported in Table 1.

---

[3] In a complete data cluster, the number of observations would be 195.

[4] We use two different subscripts in our article. Subscript j refers to the (five) classes of species, whereas h refers to the species alone (the total number of species varies from 2 to 14 in our sample).

[5] Unit price refers to the market price depending on species, diameter and quality. In the empirical model, we use the average unit price, i.e., the unit price for one species in one compartment.

[6] In reality, a more in-depth ecological study would take pH, moisture and altitude into account. There is actually no significant variation in altitude since all forests observed in our sample are located at altitudes below 500 meters.

Variable	Definition	Unit	Mean	Standard deviation	Minimum	Maximum
Dependent						
RICH	Richness index		7.80	2.84	2.00	14.00
SHANV	Shannon index (in volume)		1.17	0.42	0.07	2.16
SHANN	Shannon index (in number)		1.39	0.41	0.00	1.98
Y	Timber harvest	m^3/ha/year	1.02	1.86	0.03	15.03
Independent						
P	Timber price	euros/m^3	31.34	31.53	3.00	170.20
DIAM	Tree diameter	centimeters	30.40	12.17	9.50	58.93
QUAL%	Percentage of quality		0.25	0.16	0.00	0.62
INV	Stock inventory per species	m^3/ha	48.23	72.14	0.75	667.52
INVD	Stock inventory (sum of species)	m^3/ha	131.55	81.71	59.00	677.00
VOLINCR	Volume increment (of stock)		4.31	2.27	1.90	17.90
	Type of owners (Dummies)					
DUMO1	Institution		0.1078			
DUMO2	Forest owner		0.3333			
DUMO3	Group of owners		0.4216			
DUMO4	Joint ownership		0.1373			
	Manager (Dummies)					
DUME1			0.0686			
DUME2			0.1667			
DUME3			0.2157			
DUME4			0.0980			
DUME5			0.1176			
DUME6			0.0490			
DUME7			0.0490			
DUME8			0.0392			
DUME9			0.0392			
DUME10			0.1568			
	Ecological conditions (Dummies)					
ST1	Calcareous		0.2059			
ST2	Calcareous clay		0.0490			
ST3	Silt and clay		0.1961			
ST4	Hydromorphic		0.3333			
ST5	Sand		0.1176			
ST6	Sandstone		0.0784			
ST7	Acid		0.0196			

Table 1. Descriptive statistics, 102 observations

3.2 Estimation results

We first estimate the tree diversity demand equation (6). The diversity of tree species is observed for each forest compartment and is cluster-invariant. Since some explanatory variables vary according to the forest compartment as well as to the species (such as the price), the diversity equation is estimated by a (between-type) OLS method. All variables can be considered as exogenous in this estimation (at least, on the short term). In particular, the price is determined by the market. Hence, there can be no doubt about the direction of the cause-effect relationship. For example, it is the timber price that explains the tree diversity in a compartment and not vice versa.

As mentioned above, there are three different indices to proxy diversity. Three regressions were successfully run with the three different indices as dependent variables. The estimated coefficients are similar. However, the goodness of fit as well as the significance of parameters are better with the logarithm of the number of species (i.e., the richness index). The richness varies as soon as an individual of a new species is added or removed. Shannon indices are preferred by ecologists because they take the richness as well as the distribution of species into account at the same time. However, according to the managers, taking biodiversity into account tends to favour minority species. Estimation results are presented in Table 2.

Variable	Coefficient	s.e	Variable	Coefficient	s.e.
Dependent variable: $\ln z = \ln RICH$					
Constant	-4.1585**	1.7642	$DUME6$	-0.3070***	0.0837
$\ln p$	0.3110*	0.1781	$DUME8$	1.2389***	0.2417
$\ln DIAM$	0.7357**	0.2985	$ST1$	2.4252***	0.4362
$QUAL\%$	-2.4432***	0.6085	$ST2$	2.3281***	0.3816
$\ln INVD$	0.0645	0.2351	$ST3$	2.2615***	0.3882
$VOLINCR$	0.4216***	0.1398	$ST4$	1.8120***	0.4080
$DUME2$	0.3974***	0.0970	$ST5$	1.7468***	0.4016
$DUME3$	0.1129	0.1259	$ST6$	1.8774***	0.4155

Notes: $n = 102, N = 39$. Adjusted $R^2 = 0.602$. Heteroscedasticity-consistent s.e.

***: significant at 1%, **: significant at 5%, *: significant at 10%.

Table 2. Demand estimation - OLS method

The overall performance of the demand equation is good since the adjusted R^2 is equal to 0.602. The estimated parameters are all significantly different from zero, except for the stock inventory (i.e., the standing timber per ha) and a dummy variable that proxies a forest expert. However, other variables related to the state or trend of forest capital such as the average diameter of trees, the share of qualitative stand wood and the volume increment of forest are significant in our model. In particular, the negative sign for the coefficient associated with the percentage of quality ($QUAL\%$) has an interesting interpretation. Forests with the highest percentage of quality correspond to ones with the lowest diversity. Since a high percentage of quality increases the revenues over time, this result would mean that in this case, species diversity is less favoured, showing a trade-off between quality and diversity. Moreover, the coefficient associated with the variable $DIAM$ is significantly positive. In order to favour diversity, some trees were harvested early to diminish natural competition between species. As expected, the site context has a significant impact on the diversity. Coefficients associated with dummies from $ST1$ to $ST6$ are all significant with positive signs (decreasing from 2.43 to 1.88, respectively) with respect to acid soils ($ST7$), confirming a decrease in richness when the context is acid. Furthermore, the estimated coefficients allow a classification of the site conditions that is in agreement with the observed ecological link between the chemical characteristics of the soil and tree (and flora) diversity.

Some forest managers have a significant positive impact on tree diversity, while other ones have a negative impact that supports a short-term view. The variables for the type of forest owner have been removed because their coefficients were not significantly different from zero. The unit price has a significant and positive influence on the diversity. Its value (0.3110) means that a 10% decrease in timber price implies a 3.11% decrease in tree diversity. This result highlights the effect of timber price on the abandonment of species. For example, in the ecological context where diversity is the highest (14 species in our sample), a 23% decrease in price could lead to the loss of one species. Unit prices for timber are exogenous. However, average unit price (for one species in one compartment) can vary according to the distribution in the stand with respect to its quality and its size. The forest owner can therefore adjust his/her revenues by acting on these variables. One of the principles in uneven-aged forest management is to concentrate the volume increment on the high-quality trees. Hence, low-quality trees are progressively cut and, at the same time, the unit price of standing timber as well as that of harvested timber increase. Once this unit price has increased, forest managers and owners are more inclined to maintain the minority species. The objective is to reduce economic risks by finding an optimal distribution among the different species.

Using the estimates of the demand equation, the fitted value of diversity was computed and used as an explanatory variable in the timber supply equation (5). The use of generated regressors may produce non-consistent estimated standard errors. This is why a vector of regressors was used that includes some or all exogenous variables already in the first regression (Pagan, 1984). This second-step OLS leads, in fact, to a two-stage least squares procedure since the regressors are variables used in the first-step estimation (of the demand equation), and gives correct standard errors. Because the predicted diversity $\ln \widehat{RICH}$ can be approximated by a linear function of the explanatory variables in the demand equation and leads to a problem of collinearity, several exclusion restrictions were used in the supply equation. Some variables that do not appear to be significant to explain harvesting have thus been excluded, including the volume increment of stock ($VOLINCR$) and some dummies that proxy the forest manager. Finally, this estimation procedure is implemented with a robust variance-covariance matrix.

Within and GLS methods (for FE and RE models, respectively) are implemented as described in the econometric method section. They are also conducted in two steps like the OLS method. A Hausman test was then computed to check for the exogeneity of explanatory variables. The value of the statistic is 3.217 (with a P-value of 0.5222) and is below the $\chi^2(4)$ critical value at the 1% level. This result confirms the exogeneity of variables. Hence, the GLS method is the best adapted here for dealing with the cluster feature of our sample. R^2 is equal to 0.606

and indicates a good fitting of our model. Estimation results of (second-step) OLS, Within and GLS methods are reported in Table 3.

Variable	OLS (Pooled)		Within (FE - Fixed Effects)		GLS (RE - Random Effects)	
	Coef.	Robust s.e.	Coef.	Robust s.e.	Coef.	s.e.
Dependent variable: ln y						
Constant	-1.2893*	0.7816	-3.3204***	0.2433	-1.3713**	0.5866
ln p	0.5735***	0.0955	0.3467***	0.1089	0.4659***	0.0699
ln $DIAM$	-0.0368***	0.0089	-0.0424***	0.0072	-0.0387***	0.0066
$QUAL$%	-1.4822**	0.6484	-1.1813	0.7142	-1.3775***	0.4515
ln INV	0.7178***	0.0771	0.9054***	0.0756	0.8085***	0.0630
ln $\widehat{RICH}$	-0.8651**	0.3456	–		-0.7797***	0.2610
$DUME2$	-0.8210*	0.4226	–		-1.0129***	0.2387
$DUME3$	-0.3628	0.2651	–		-0.4820**	0.2119
$ST4$	-0.5238**	0.2433	–		-0.5008**	0.1980
$ST5$	-0.8125**	0.3761	–		-0.7456**	0.3018
$\hat{\sigma}_{\epsilon}^{2}$					0.5065	
$\hat{\sigma}_{\mu}^{2}$					0.2861	
Hausman test (P-value)					3.217 (0.5222)	
R^2	0.613		0.460		0.606	

Notes : n=102, N=39. ***: significant at 1%, **: significant at 5%, *: significant at 10%. Robust s.e. for OLS and FE estimation are respectively computed following Pepper (2002) and Arellano (1987).

Table 3. Supply estimation - Cluster-sample econometric methods

As explained above, OLS is less efficient than GLS since it does not fully take the cluster feature of our sample into account, even if a robust variance-covariance matrix makes it possible to alleviate this problem. OLS coefficients are rather similar to those estimated by specific cluster methods. However, some interest coefficients such as those associated with price and diversity are slightly overestimated. For example, the coefficient associated with the price is 0.57 with OLS, compared to 0.43-0.47 with GLS. For the diversity, it is equal to 0.87, compared to 0.75-0.78 (in absolute value).

The coefficients associated with the variables $QUAL$% and $DIAM$ are significantly negative (with estimates of -1.38 and -0.04, respectively). This means that high-quality trees with big diameters are harvested to be sold. Hence, the actual standing timber is characterised by a lower percentage of quality and a lower average diameter. Moreover, it is not surprising to see that whereas forest managers have a positive impact on tree diversity, this is not the case for timber harvest.

Results also show a positive and significant impact of both timber inventory and unit price. As expected, timber harvest increases with the standing volume of trees. The coefficient associated with the price (or price elasticity of timber supply) is estimated at 0.47, meaning that a 10% increase in price implies a 4.7% increase in harvesting.

The diversity is negatively and significantly correlated to the timber harvest, all things being equal. The estimated coefficient can be directly interpreted as a measure of substitution between tree diversity and the volume of timber harvested. The point estimate is equal to -0.78. This value is rather high. However, based on the standard error estimate, we can reject the hypothesis of a unitary elasticity substitution. An explanation for this negative sign is that when the site context is acid, the forest manager cannot influence the unit timber price interval per species. In this case, under acid soil conditions, the forest manager can only act on timber volume. On the contrary, the basic context allows for a greater variety of species. However, in order to favour all species, the forest manager cannot increase the standing volume and in some cases, may be forced to reduce it. Hence, the forest stock is low on the long term and this trend leads to a lower timber harvest.

4. Conclusion

In this study, a household production approach was used to model the behaviour of the NIPF owner in order to derive the structural econometric equations of timber supply and diversity demand and to estimate substitution and price elasticities. In the empirical application, a definition of diversity was chosen solely on the basis of the number of tree species. This diversity is simple to calculate and positively correlated with the diversity in flora and fauna. Moreover, the richness of data related to harvested species and the cluster-sample methods used in this context make it possible to deal with heterogeneity and variability within clusters. In addition, Within and GLS estimation methods make it possible to test for the possible endogeneity problem of some variables.

This study revealed that diversity demand and timber supply are negatively linked, meaning that an increase in tree diversity will lead to a decrease in timber harvesting. This result confirms that these two forest outputs are substitutes. Estimation also shows that timber price and tree diversity evolve in the same direction: the positive and significant coefficient associated with the timber price in the demand equation indicates that a price decrease has a negative effect on diversity. This result is certainly the consequence of the characteristics of uneven-aged forests and the strategies used to manage them. This could be explained by the fact that a part of the diversity not only procures some satisfaction for the forest owner, but that the price paid for this diversity is a decrease in timber production. Management strategies should therefore be aimed at finding a trade-off between timber production and tree diversity in a given ecological context.

5. References

Amacher, G. S., Conway, M. C. & Sullivan, J. (2003). Econometric analysis of forest landowners: is there anything left to do?, *Journal of Forest Economics* 9(2): 137–164.

Arellano, M. (1987). Computing robust standard errors for within-groups estimators, *Oxford Bulletin of Economics and Statistics* 49(4): 431–434.

Baltagi, B. H. & Chang, Y.-J. (1994). Incomplete panels : A comparative study of alternative estimators for the unbalanced one-way error component regression model, *Journal of Econometrics* 62(2): 67–89.

Barbier, S., Gosselin, F. & Balandier, P. (2008). Influence of tree species on understory vegetation diversity and mechanisms involved - a critical review for temperate and boreal forests, *Forest Ecology and Management* 254(1): 1–15.

Binkley, M. (1981). *Timber Supply from Non-Industrial Forests: A Microeconometric Analysis of Landowner Behavior*, Yale University Press, New Haven, CT.

Bruciamacchie, M. & de Turckheim, B. (2005). *La Futaie Irrégulière: Théorie et Pratique de la Sylviculture Irrégulière, Continue et Proche de la Nature*, Édisud, France.

Ehrlich, P. R. & Raven, P. H. (1964). Butterflies and Plants: A Study in Coevolution, *Evolution* 18(4): 586–608.

Greatorex-Davies, J. N., Sparks, T. H., Hall, M. & Marrs, R. H. (1993). The influence of shade on butterflies in rides of coniferised lowland woods in southern england and implications for conservation management, *Biological Conservation* 63(1): 31–41.

Hyberg, B. T. & Holthausen, D. M. (1989). The behavior of nonindustrial private forest landowners, *Canadian Journal of Forest Research* 19(8): 1014–1023.

Lähde, E., Laiho, O., Norokorpi, Y. & Saksa, T. (1999). Stand structure as the basis of diversity index.

Max, W. & Lehman, D. E. (1988). An behavioral model of timber supply, *Journal of Environmental Economics and Management* 15(1): 71–86.

Mayle, B. A. (1990). A biological basis for bat conservation in british woodlands - a review, *Mammal Review* 20(4): 159–195.

McDermott, M. E. & Wood, P. B. (2009). Short- and long-term implications of clearcut and two-age silviculture for conservation of breeding forest birds in the central Appalachians, USA, *Biological Conservation* 142(1): 212–220.

Moulton, B. R. (1986). Random group effects and the precision of regression estimates, *Journal of Econometrics* 32(3): 385–397.

Newman, D. & Wear, D. (1993). Production economics of private forestry: A comparison of industrial and non-industrial forest owners, *American Journal of Agricultural Economics* 75: 674–684.

Pagan, A. (1984). Econometric issues in the analysis of regressions with generated regressors, *International Economic Review* 25(1): 221–247.

Pattanayak, S. K., Abt, K. L. & Holmes, T. P. (2003). Timber and amenities on nonindustrial private forest land. in E. Sills, K. Abt (Eds.), Forests in a Market Economy, Kluwer Academic Publishers, Dordrecht.

Pattanayak, S. K., Murray, B. C. & Abt, R. C. (2002). How joint is joint forest production? an econometric analysis of timber supply conditional on endogenous amenity values, *Forest Science* 48(3): 479–491.

Pepper, J. V. (2002). Robust inferences from random clustered samples: An application using data from the panel study of income dynamics, *Economics Letters* 75(3): 341–345.

Schuldt, A., Fahrenholz, N., Brauns, M., Migge-Kleian, S., Platner, C. & Schaefer, M. (2008). Communities of ground-living spiders in deciduous forests: Does tree species diversity matter?, *Biodiversity and Conservation* 17(5): 1267–1284.

Swamy, P. A. V. B. & Arora, S. S. (1972). The exact finite sample properties of the estimators of coefficients in the error components regression models, *Econometrica* 40(2): 261–275.

Wooldridge, J. M. (2003). Cluster-sample methods in applied econometrics, *American Economic Review* 93(2): 133–138.

6

Models to Implement a Sustainable Forest Management – An Overview of the ModisPinaster Model

Teresa Fonseca[1], Bernard Parresol[2],
Carlos Marques[1] and François de Coligny[3]
[1]Department of Forest Sciences and Landscape Architecture (CIFAP),
University of Trás-os-Montes e Alto Douro
[2]USDA Forest Service, Southern Research Station
[3]Institut National de la Recherche Agronomique - botAnique et bioInforMatique de
l'Architecture des Plantes, INRA-AMAP
[1]Portugal
[2]USA
[3]France

1. Introduction

For a long period, practical recommendations for forest management were based upon experience gained through trial and error experimentation, observation and an understanding of density effects on tree growth within the stand. As stated by Zeide (2008), the limitations of the traditional empirical approach coupled with improvements on modelling efforts led to a change of procedures from forestry to forest science, this being defined by the author, as a new development relying on reasoning to produce the optimal system of forest management aimed at satisfying human needs and preserving nature at the same time (though not at the same place).

Nowadays, the use of mathematical models for tree and stand growth dynamics is the recommended scientific approach to test for alternative management options under a Sustainable Forest Management (SFM) concept and to help solve practical problems such as the appropriate range of stand densities, the thinning prescriptions and rotation ages that allow for a given goal. Assessment of volume and biomass growth, for a given period, or of yield and carbon stock at a point in time becomes a straightforward procedure as long as there are proper equations available, for the species and region of study.

Central to the successful implementation of research findings of sustainable forest management is their efficient transfer from the researcher to the manager (Farrell et al., 2000). In this context, there is a strong need for easily accessible programs to run various and numerous simulations in a convenient and flexible way. There are different possible approaches to build a simulation system, each having advantages and drawbacks. One is to build a specific tool for each model. Development can be fast when the objectives of the model are well defined, its structure remains simple and there is no need for complex outputs and interfaces. This approach nevertheless results in building many prototypes which are generally not very flexible and are difficult to reuse. A second approach is to build one tool around a reusable model and adapt it to different species and situations by changing model parameters. The main drawback remains the limitation to one model with little possible modification. A more interesting option is given by Capsis (Computer-Aided Projection of Strategies In Silviculture, http://www.inra.fr/capsis). The Capsis software is a domain specific tool with a common methodology, but accepting models with different data structures, simulation steps and evolution methods.

This chapter will focus on a forest model developed for maritime pine (*Pinus pinaster* Ait.), the ModisPinaster model (Fonseca, 2004), as a supporting tool for Sustainable Forest Management that is freely available for use in the user-friendly Capsis platform. ModisPinaster (Model with Diameter Distribution for *P. Pinaster*) is a dynamic growth and yield (G&Y) model that applies to pure maritime pine stands. It is constituted by several components allowing the simulation of stand evolution through the rotation period and the simulation of interventions such as thinning and clear-cut. In the mortality component, abiotic and biotic variables are used to determine forest vulnerability to damages from wind and snow. This feature is invaluable in a climate change adaptation scenario. The level of detail of the output is the diameter class, with the diameter distributions being recovered by the 4-parameters Johnson S_B distribution (Johnson, 1949; Fonseca et al., 2009; Parresol et al., 2010). The model can be downloaded from the CAPSIS simulation platform web site.

This chapter has the following structure. Section 2, gives an overview of forest models that have been proposed for maritime pine (2.1), followed by a description of the Capsis platform (2.2) and its current uses. Section 3 is devoted to the description of the structure (3.1) and subcomponent models of the ModisPinaster model (3.2). A portrayal of ModisPinaster simulation capabilities within the Capsis interface is depicted in (3.3). Section 4 presents an example of simulation of three management scenarios for the species, using the CAPSIS environment. One scenario follows the traditional management guidelines in the study area. The second scenario follows density management criteria according to the self-thinning line theory. A third scenario provides a simulation that is compatible with the biodiversity promotion, under a SFM policy. Concluding remarks are presented in Section 5.

2. Use of forest models as a supporting tool for SFM

The sustainable management of the forests has been seen, for a long time, as a sustained yield of wood supply. Thus, it is not unexpected that until recently the great emphasis in the forest research domain has been towards stand volume predictions. Estimates on timber volume production originally come from spacing and thinning trials. The field experiments led to the creation of the first generation of yield tables, by German scientists, in the late 18th to the middle of the 19th century. From the experimental tables to the present G&Y models, different types were developed; although their main uses still are for timber management purposes.

The onset of the multi-functional forest paradigm caused the development of models for other purposes such as: management of non-wood products, the promotion of biodiversity, increasing the social benefits and aesthetic demands. For instance, according to the EU commission study (Nieto & Alexander, 2010), in Europe, 11% of the saproxylic beetle species are currently threatened. The main threat, relates to the loss and decline of their habitat either in relation to logging and wood harvesting in forests or due to a general decline in veteran trees throughout the landscape. Management can help to conserve the biodiversity if done in a sustainable way (e.g. leaving dead wood material in the forest) and promoting the existence of older trees. Attention has also been focused on modelling natural disturbances as they can seriously affect timber production and other forest benefits. It is worthwhile to say that although simulations help to provide management guidelines for the forests, nature is not a virtual forest. Unexpected results might occur in a real forest under a real management process. Critical evaluation of results and adaptive management procedures that take risk into account are therefore advocated when using models for forest growth simulation purposes.

Contemporary studies in modelling extend to the use of the physiologic process based models. Nevertheless, process based models are generally not considered feasible for predicting G&Y under a SFM, as they require a great number of variables and parameters. Some of these variables (e.g. daily meteorological data, radiation absorption, transpiration rate) are hardly accessible or cannot be measured at all and their values have to be guessed, and the models are not capable of providing adequate predictions of tree growth (Zeide, 2008).

An overview of the model approaches for management of European forests is presented by Pretzsch et al. (2008), while a review of models employed to deal with the complexities associated with natural disturbance processes can be seen in Seidl et al. (2011).

Pretzsch et al. (2008), state that the objectives and structure of a model reflect the state of the art of the respective research area at the time, and document the contemporary approach to forest growth prediction. This does not mean that a new model is a better model than the previous existing ones. The selection of one model instead of a past one is strongly dependent on the reliability of the model and on the accuracy of the estimates, which are both dependent on the quality of the supporting data. From the model user's point of view, other useful features are the minimum input requirements; the ability for allowing simulations for diverse combinations of the state variables; and the ease of use. With the development of modelling software, this has become straightforward.

2.1 Available models to help for a SFM of *Pinus pinaster*

The maritime pine (*Pinus pinaster* Ait.), originally from the Mediterranean Basin, is an important conifer species in Portugal, Spain and France occupying an area greater than 3 500 thousand hectares (885 000 ha in Portugal, 1 684 000 ha in Spain and 1 100 000 ha in France). The first evidence of the species in Portugal, dates from the Pleistocenic, about 33 000 years ago. Now it is the leading softwood species in the country covering 27% of the mainland forested area. The major continuous cover is located in the central part of the country, in Mata Nacional de Leiria (MNL) and in the north of the country, in the Tâmega Valley region (TVR). The main uses of the species are related to wood for timber and pulp and to a lesser extent in resin production. On the poor sites it is used in afforestation programs for soil protection. The rotation age usually ranges from 40 to 50 years, although higher rotation ages do occur namely when aiming at high target diameters. To help with the management of the species, several G&Y models have been proposed. The earliest refers to the stand tables by Santos Hall (1931) for the even-aged stands in the MNL in Portugal. The tables by Echeverría & de Pedro (1948) for the pine stands in Pontevedra (Galicia), and the tables by Décourt & Lemoine (1969), for the pine stands in the South-West region, are the first references in Spain and in France, respectively.

Traditionally, the stand tables were based on a reference stand, originally of normal density or fully stocked, and site index, following predetermined average silviculture guidelines. The development of improved analysis methods has allowed for new types of models, which are not restricted to tabular forms or to fixed densities. For Portugal, the most recent include the diameter distribution models PBRAVO (Páscoa, 1987) and ModisPinaster (Fonseca, 2004), Dryads (Gonçalves, 2003) for mixed stands, and PBIRROL for uneven-aged structures (Alegria, 2003). For Spain, there are PINASTER (Soalleiro et al., 1994; Soalleiro, 1995) for even-aged stands in Galicia and the model developed by Diéguez-Aranda et al. (2009), included in the GesMO platform. Orois & Soalleiro (2002) proposed a model that applies to mixed stands. The platform SIMANFOR (Bravo et al., 2010), not being a model, integrates a set of modules for simulating and projecting stand conditions in Central Spain.

Available G&Y models for maritime pine in France, are one whole-stand model named PP1 (Lemoine, 1991) and two distance independent models, Afocelpp (Najar, 1999), and PP3 by B. Lemoine, P. Dreyfus and C. Meredieu (derived from Lemoine, 1991; Salas-Gonzalez et al., 2001) (http://www.inra.fr/capsis/models). The latter model presents several functionalities such as a dead wood estimation (Brin et al., 2008), windthrow risk through a connection to ForestGales (Cucchi et al., 2005), and wood quality assessment (Bouffier et al., 2009). The three French models and ModisPinaster are currently integrated in the Capsis platform.

2.2 The CAPSIS 4 platform

Software development can be very time-consuming and expensive. The Capsis project has undergone continuous development in France since 1994 with the aim to simulate the consequences of silvicultural treatments based on scientific knowledge, and to build an integration platform for forestry growth and yield models. One of the objectives of Capsis is to share this effort by organising the work around a small number of software developers, who concentrate on the technical aspects of the software and common tools, and modellers who concentrate on specific modules related to the scientific core of their models. Capsis relies on the JAVA environment. This choice of an object-oriented language promotes easier adaptation of common ancestor objects by the modellers, as well as modularity. Capsis has an open software architecture around a stable kernel, augmented with applicative and technical libraries. Different models are integrated in it as many modules, and various tools can be added at any time within flexible extensions. This "platform" runs either in interactive context to explore possibilities or in batch mode to run long or repetitive simulations (Dufour-Kowalski et al., 2012). Initially developed for forest modellers, the range of Capsis end-users very quickly expanded to a large number of stakeholders (Figure 1). It is used in an increasing number of applications for forest management and training. Stakeholder aims are now: to contribute to the development of models and test their sensitivity to model parameters by simulating managers' actions, to share tools and methods, to compare results of different models, to transfer models to managers and to develop training material.

Every component developed inside Capsis, except the Capsis modules (i.e. the model implementations) can be freely distributed under a free license (Lesser General Public Licence), meaning that the core application, including all the extensions, can be used by anyone.

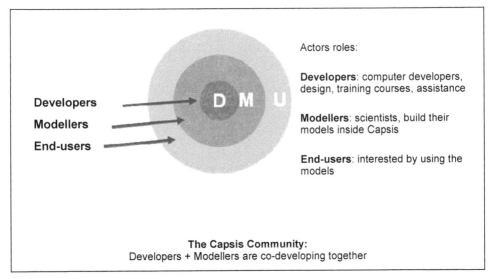

Actors roles:

Developers: computer developers, design, training courses, assistance

Modellers: scientists, build their models inside Capsis

End-users: interested by using the models

Developers

Modellers

End-users

The Capsis Community:
Developers + Modellers are co-developing together

Fig. 1. The Capsis project organization.

Concerning the modules (i.e. the models), the authors decide on the license they wish, free or not, and choose the way to distribute them outside the community. This framework relies on mutual confidence and favours multiple public and private partnerships. The current release of Capsis (Capsis 4) now contains more than 50 forest growth or dynamics models of different types: distance-independent tree models and individual tree models, as well as mixed models, developed by modellers worldwide (http:// www.inra.fr/capsis/models). In addition, models within Capsis can be connected with other software (GIS, visualisation, architectural models, de Coligny, 2007). The potentialities of Capsis enhance the use of forest models for SFM through the ease of sharing the models with forest managers without charges and permitting the analysis of different scenarios. Simulations are easy to run and users can utilize and test different silvicultural scenarios for a sustainable forest management. It is used in an increasing number of applications for forest management and training. Capsis is particularly useful in situations where observation and experimentation is difficult. Many local and regional French National Forest Service offices have used the Capsis software to help define management operations for implementation by the field services (Meredieu et al., 2009). For example the silviculture handbook for the French northern Alps, applied to both public and private forests, is based on Capsis simulations, especially for mixed fir-spruce forests (Gauquelin & Courbaud, 2006).

3. A case study: the ModisPinaster model

Data used in the model development come from a large database on maritime pine (Data_Pinaster) created and maintained over the last two decades at the Department of Forest Sciences and Landscape Architecture of the University of Trás-os-Montes e Alto Douro. Data come from temporary and permanent plots and were collected in northern Portugal, more precisely in stands located in the Tâmega Valley (latitude range: 41° 15′N –

41°52′N; longitude range: 7° 20′ W – 8° 00′ W). The model addresses forest growth and yield, risks (wind related) and management procedures such as thinning and harvesting. Since its development, efforts have been made to promote its dissemination to potential users and to allow for a more effective use under a SFM vision. The implementation of ModisPinaster within the Capsis platform has improved its use as a tool for sustainable management of maritime pine forests. Details are given in the following sections.

3.1 Description of Modispinaster

ModisPinaster is constituted by six components: (i) dominant height growth; (ii) basal area growth; (iii) tree mortality; (iv) diameter distribution; (v) thinning algorithm and (vi) output functions for volume, biomass and carbon content assessment by diameter classes. The relationship among the components is shown in Figure 2.

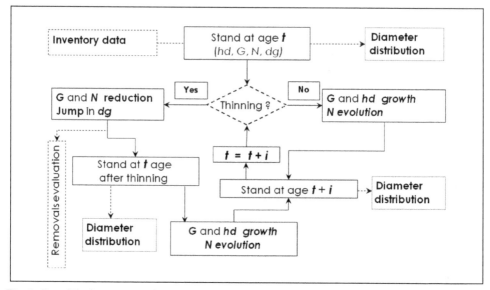

Fig. 2. Simplified structure of ModisPinaster.

The model initiates from a calibration point that requires data variables easily obtained from current inventories:

- Stand variables: stand age (t, yrs), the average height (m) of the 100 thickest trees per ha (hd, m) or the site index value (SI, m, base age 35 years), basal area (G, m² ha⁻¹), number of trees per hectare (N, trees ha⁻¹) and the average diameter of the dominant trees (dd, cm).
- Site variables: terrain slope (Inc, °.) and terrain direction (Exp, °.).
- Stand nature: specified as a qualitative variable (homogeneous or heterogeneous in terms of the uniformity of trees' age) or assessed through the diameter distribution in terms of the number of classes (5-cm wide) and the standard deviation of the diameters (sd, cm).
- Management variables (optional): number of trees recently cut (if any) (N_t, trees ha⁻¹).
- Historical details on tree mortality (optional): presence or absence of dead trees (0/1).

Other optional variables include: the median ($d_{0.50}$, cm), the average ($\bar{d}$, cm) and the minimum value of the diameter distribution (d_{min}, cm).

The input data coupled with the model components allow representing the stand growth and the management practices that are typical to the species, including the simulation of mechanical and selective thinnings and clear-felling. The maximum age allowed for the rotation term is 65 years. The minimum scale level admitted for prediction is the year.

Dominant height growth, for a given site index, is estimated using Marques (1987) model (equation 1). Site index value is calculated from Marques (1987, 1991) SI model (equation 2).

$$\widehat{hd} = e^{4.04764 - 8.75819t^{-0.56087}} + 1.19874\left(1 - e^{-0.081t}\right)^{2.99578}\left(SI - 17.38\right) \tag{1}$$

$$SI = 17.38 - \left(e^{4.04764 - 8.75819t^{-0.56087}} - hd\right) \times$$
$$\left(0.865685 - 0.00804747t + 0.0009994305t^2 - 0.0000187066t^3\right) \tag{2}$$

Basal area at the projection age is estimated with equation 3. The growth model was originally presented in Svetz & Zeide (1996), and was refitted by T. Fonseca, after Fonseca (2004), using the Data_Pinaster dataset.

$$\hat{G}_2 = \left[G_1^{0.4090} + 7.4949e^{-0.0333t_1}\left(1 - e^{-0.0333(t_2 - t_1)}\right)\right]^{1/0.4090}$$
$$e^{-0.8427(N_1 - N_2)/N_2} \tag{3}$$

In equation (1), SI refers to site index, defined as the stand dominant height (hd) at the reference age of 35 years whereas, in equation (3), G_i and N_i refer to the stand basal area and to the number of trees at age t_i, respectively (i = 1, 2 for actual and projection age, respectively). The other variables in equations 1-3 were already defined.

Evaluation of tree mortality is a two-phase process. In the first phase the model estimates the probability of mortality to occur during the projection period. In a second phase the number of survival trees is calculated for the projection age and then it is adjusted by the probability of mortality to occur.

Probability of mortality is predicted by two equations developed by Fonseca (2004), according to the major influences: wind (equation 4) and other causes (equation 5), these being mainly related to competition effects.

$$\hat{p}_1 = \left[1 + \exp\left(-\left(\begin{array}{c}-30.4753 + 0.3725Qhdc + 21.1705RS_b \times BRS \\ +4.2303BSExp + 0.1758Incl \times BF_1 + 5.5347BF_2\end{array}\right)\right)\right]^{-1} \tag{4}$$

$$\hat{p}_{OC} = \left[1 + \exp\left(-\left(\begin{array}{c}-8.5235 - 0.3822N_t/100 + \\ (t_2 - t_1)(2.1449 + 1.5768BMA \times C + 39.2942BAMPD/dg_{MAX})\end{array}\right)\right)\right]^{-1} \tag{5}$$

In equation 4 and equation 5, the variable $Qhdc$ and the binaries BF_1 and BF_2 are related to the stability of the stand; the variable RS_b refer to relative spacing before thinning with BRS being a binary that makes a distinction of the average space conditions between the trees for the current stand; dg_{MAX} is the maximum values of tree diameter allowed according to the self-thinning line for the species (Luis & Fonseca, 2004) and *BAMPD* is a binary variable used to differentiate the stands according to the proximity to the self-thinning line. These variables and the ones related to the occurrence of recent mortality (*BMA*) and to the intrinsic risk of damages occurrence, due to the terrain direction (*BSExp*) are defined as follows:

$Qhdc$	=	$100 \times (hd - 1.30m)/dd$
BF_1	=	1 if $Qhdc > 48$; $BF_1 = 0$, otherwise.
BF_2	=	1 if $100 \times hd/dd \leq 54$; $BF_2 = 0$, otherwise.
RS	=	$100/(\sqrt{N} \times hd)$
RS_b	=	$100/(\sqrt{N_b} \times hd)$
BRS	=	1 if $RS \leq 0.20$; $BRS = 0$, otherwise.
C	=	1 for recent thinning; $C = 0$, otherwise.
dg_{MAX}	=	$25\,(1859/N)^{1/1.897}$
$BAMPD$	=	1 if $(dg_{MAX} - dg) \leq 17.5cm$; $BAMPD = 0$, otherwise.
BMA	=	1 for recent mortality; $BMA = 0$, otherwise.
$BSExp$	=	1 if direction (°) belongs to $]60, 120] \cup]180, 240] \cup]300, 360]$; $BSExp = 0$, otherwise.

A join probability of mortality for the period t_2-t_1 is estimated as

$$\hat{p} = \hat{p}_1 + \hat{p}_{OC} - \hat{p}_1\hat{p}_{OC} \tag{6}$$

An initial assessment of the living trees at projection age t_2 is given by the survival model (equation 7), developed after Huang et al. (2001).

$$\hat{N}_2 = N_1 \left(\frac{1 + \exp\left[-5.2560293 + 1.81990161\ln(1+t_1) - 0.1532847SI + 0.86246466BE\right]}{1 + \exp\left[-5.2560293 + 1.81990161\ln(1+t_2) - 0.1532847SI + 0.86246466BE\right]} \right) \tag{7}$$

In equation 7 BE is a binary variable that characterizes the stand horizontal structure based on the heterogeneity of tree diameters. The stand is homogeneous in composition ($BE = 0$) for diameter distributions not exceeding 25 cm in range and 5.5 cm in standard deviation; otherwise it is heterogeneous ($BE = 1$).

The number of trees is adjusted according to equation 8.

$$\hat{N}_{2aj} = N_1 - \hat{p}(N_1 - \hat{N}_2) \tag{8}$$

ModisPinaster is a distribution model that presents output information at the detailed level of the diameter class. The diameter distribution is modelled by Johnson's S_B (Johnson, 1949), (equation 9).

$$f(d) = \frac{\delta\lambda}{\sqrt{2\pi}(d-\xi)(\xi+\lambda-d)} \exp\left(-\frac{1}{2}\left[\gamma + \delta\ln\left(\frac{d-\xi}{\xi+\lambda-d}\right)\right]^2\right) \quad \xi < d < \xi + \lambda \tag{9}$$

$$= 0, \text{ otherwise}$$

where λ, $\delta > 0$, $-\infty < \xi < \infty$, $-\infty < \gamma < \infty$; λ is a range; ξ is a location parameter (lower bound), δ and γ are shape parameters, $\gamma = 0$ indicating symmetry.

The algorithm used to incorporate the S_B distribution in ModisPinaster was based on the parameter recovery method in combination with the parameter prediction proposed by Parresol (2003). Briefly, in his approach, Parresol assumed the minimum location parameter was pre-specified (set to 0.8 of minimum diameter in ModisPinaster). The range and two shape parameters were then recovered from the median and the first two noncentral moments of the diameter distribution (average diameter and quadratic mean diameter). A complete SAS code for the procedure is available in Parresol et al. (2010). The evolution of the stand structure in terms of diameter class distribution is provided for each year of the simulation and whenever an intervention is simulated.

The procedure to represent the stand structure after a thinning requires previous information on diameter distribution (actual or simulated using the S_B distribution). Trees to be removed from the diameter distribution are identified with a thinning algorithm (Alder, 1979). The procedure assumes a probability of survival to cut proportional to a tree's size, $l(F) = Fc$, with c given by N_t / N_a. The number of trees that remain in the diameter class j (N_{ja}) is then calculated as:

$$N_{ja} = N_b L \left[F(d_j)^{1/L} - F(d_{j-1})^{1/L} \right] \tag{10}$$

In equation 10, F represents the initial probability density function (PDF) and L corresponds to the proportion of the standing trees, comparing to the number of trees before thinning ($L = N_a/N_b$). The diameter distribution for the removed trees is obtained by subtraction.

The stand basal area after thinning is calculated with equation 11. This equation was refitted by T. Fonseca, after Fonseca (2004), using the Data_Pinaster dataset.

$$G_a = G_b \left(N_a / N_b \right)^{2.4979 N_b^{-0.1951}} \tag{11}$$

Stand variables after thinning are used to recalibrate stand variables at age t. That is, N becomes equal to N_a and G becomes equal to G_a.

Auxiliary functions, not presented here, were developed to allow for the estimation of the optional input variables. These include functions for the median, the average, the minimum and the maximum values of the diameter distribution, all required for the S_B recovery procedure. Additional functions appended to ModisPinaster refer to the height-diameter relationships, and to tree equations to calculate the volume and biomass content. At present, the model uses the height-diameter relationships by Almeida (1999), the volume equations from Fonte (2000), and the biomass equations from Lopes (2005).

3.2 ModisPinaster within the Capsis interface

The potentialities of Capsis enhance the use of ModisPinaster for SFM through the ease of permitting the analysis of different scenarios in a friendly environment. The extended outputs provide diverse information of stand growth and structure as well as of thinning intervention.

The simulations are easy to run and the users can utilize and test different silvicultural scenarios for a better choice under a sustainable forest management policy.

Figure 3 presents the ModisPinaster initializing scenario for a sampled stand (Stand_1) with the minimum input variables required for simulation purposes. By default, the stand is assumed to be of homogeneous structure (even-aged). For the example, a merchantability limit of a top diameter of 7 cm is specified for volume estimates. The state variables of the stands, age (t, yrs), dominant height (hd, m), number of trees (N, trees ha⁻¹), and basal area (G, m² ha⁻¹), are projected on an annual basis until the end of the growth period, using the set of equations 1-8. A portrait of the Evolution dialog and of the appended output options is depicted in Figure 4. An automatic management procedure, based on the self-thinning theory is available in the Management/Self-thinning dialog. By default, the limits specified for the Stand Density Index (SDI, Reineke, 1933; Luis & Fonseca, 2004) are with respect to a stand that grows under high values of density. The occurrence of mortality by competition is expected whenever SDI reaches the threshold of 60%. The limits can be modified by the user, according to pre-determined management guidelines.

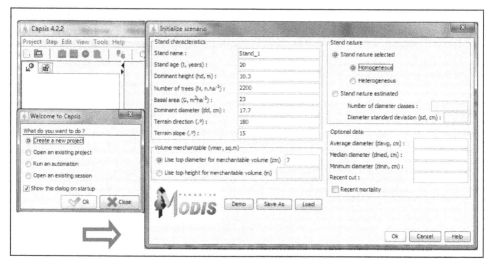

Fig. 3. The ModisPinaster input dialog in the Capsis platform.

Selected outputs shown in Figure 4 refer to a stand table for the 5-year evolution period (bottom left) and to the number of trees, per diameter class, at the initial stand age (20 yrs) and at the end of the simulation (25 yrs) (bottom right). The diameter distribution is presented for each selected scene (year) according to the methodology described in section 3.1 and detailed in Parresol et al. (2010).

Major improvements for the thinning simulation ability of ModisPinaster are available in the Capsis thinning interface (see Figure 5, to the left). Users can decide the thinning prescription based on the total number of trees to remove, or in density regulation rules, according to the stand density index (SDI) criteria based on the self-thinning theory, or according to the Wilson spacing factor, $Fw = 100N^{-0.5}hd$, where the variables N and hd had already been defined. In each of these cases, the selection of the trees to remove from the stand is made according to the algorithm of Alder, described in section 3.1, which assures a probability of a tree to survive to cut being proportional to the tree's size. Alternatively, the user can perform a selective thinning using the Capsis interactive diagram. With this option, the trees are cut by action on an interactive diameter distribution diagram. Figure 5 presents the intervention dialog with a simulation of a thinning at 20 years of age. In the example, a Wilson spacing factor of 0.23 was specified for the thinning criteria. At the bottom of the dialog, a summary of the variables N, G and dg, is presented for the initial and for the residual stand, and for the thinned material. Results obtained for a growth series of 10 years, after the thinning, are presented to the right of the figure: a stand table in an annual basis, and the disaggregation by diameter classes of the total and merchantable volumes and of carbon content, per component, at the end age.

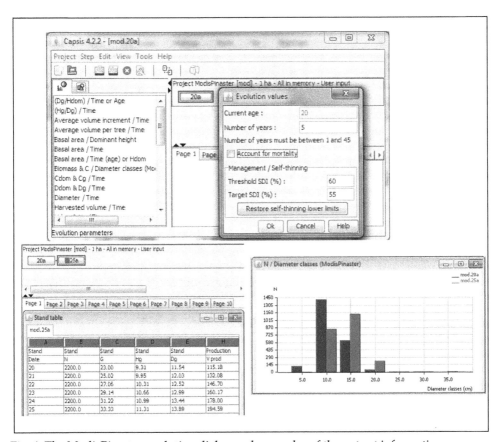

Fig. 4. The ModisPinaster evolution dialog and examples of the output information.

3.3 Analysis of different scenarios for a SFM

The majority of the maritime pine stands in Portugal are even-aged and are handled in a thinned managed regime. Density regulation is usually based on the Wilson spacing index. A typical value of Fw = 0.23 has been assumed in the maritime pine stands of the Tâmega Valley region (Moreira & Fonseca, 2002). The rotation age is defined by the age at which occurs the maximum annual increment of tree stem volume. Depending on site quality, stands attain their absolute explorability term for the volume variable when the stand reaches 35 (high quality) years to 45 (low quality) years (Marques, 1987; Moreira & Fonseca, 2002). In the area studied it is usual to set the lowest limit of stand age to harvest to 40-45 years.

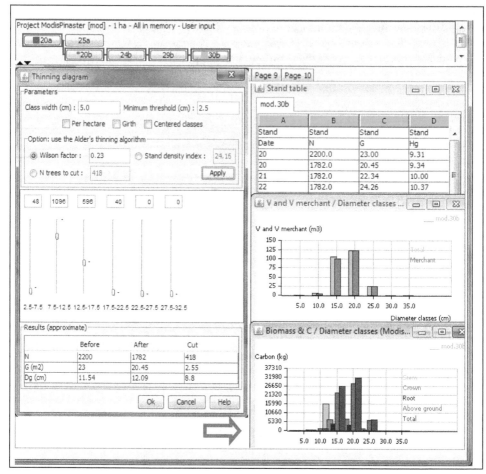

Fig. 5. The thinning dialog and some of the possible results available for ModisPinaster in the Capsis environment.

Three scenarios are proposed for comparison:

- Typical forestry guidelines (**TYF**). This scenario focuses on timber production, according to the traditional silviculture guidelines followed in the Tâmega Valley region. That is, cyclic thinning with an average silviculture compatible with a strong to moderate grade with $Fw = 0.23$; and a rotation age of 45 years.
- Low competition forestry (**LCF**). This scenario also focuses on the single purpose of timber. The management is made in accordance to the self-thinning line theory. The stand has a window for density between 25-35%, as measured by the stand density index, to keep the inter-tree competition at lower levels. The rotation age is maintained at 45 years.
- Combined objectives forestry (**COF**). Here, emphasis is done on maximizing the total volume yield and on promotion of the biodiversity. This scenario can be viewed as landowner absence where natural mortality is expected to occur. An old-stand situation is promoted. Rotation age is extended to 65 years.

The data used for the simulations refer to a 20 year old stand with the characteristics depicted in Figure 3.

For each scenario the following indicators were selected for comparison: yield in volume and in carbon in the aboveground component; products obtainable from the thinning practices (volume and average tree size). Dead wood was quantified in scenarios **TYF** and **LCF** as downed woody debris produced by thinning, considering that the tip of the trees (top diameter of 7 cm) are kept in the stand. For scenario **COF**, dead wood was quantified as the total volume of the stem for the trees that die during the rotation period. Table 1 presents the results obtained by following the current silviculture guidelines and the strategy of allowing the stand to grow in lower levels of competition. In the first case, a total of four thinnings, with a cycle of 5 years, starting at age 20, were considered until the stand reaches 35 years. Simulations of the thinning were made using the thinning dialog with specification of the target value for the Wilson factor (see Figure 5). After each thinning, the stand growth was simulated with the evolution dialog. For the **LCF** scenario, the simulation was made in an automatic mode using the facilities given in the evolution dialog (see Figure 4), setting a threshold of SDI equal to 35% and a target value of SDI equal to 25%. The first thinning occurs at the age of 24 years, a second at 30 years and a third and last one when the stand age is 37 years.

Sim.	t (yrs)	hd (m)	dg_b (cm)	N_b (trees ha⁻¹)	C_b (ton ha⁻¹)	V_b (m³ ha⁻¹)	V_t (m³ ha⁻¹)	Debris (m³ ha⁻¹)
	20	10.3	11.5	2200	28.5	115.2	14.6	3.2
	25	12.9	14.7	1782	41.1	176.9	45.6	4.2
TYF	30	14.8	18.9	1136	49.0	212.6	37.9	1.1
	35	16.5	22.6	859	58.1	255.8	37.6	0.5
	45	19.4	28.2	693	82.2	374.1		
	20	10.3	11.5	2200	28.5	115.2	-	-
	24	12.5	13.4	2200	41.0	178.0	61.8	7.6
LCF	30	14.8	18.9	1146	49.4	214.2	71.4	1.9
	37	17.1	25.2	648	57.4	252.5	79.3	0.6
	45	19.4	31.8	401	65.7	293.4		

Table 1. Characteristics of the stand, at age t (years), according to the simulation results of the management scenarios focused on timber production: typical forestry guidelines (TYF) and low competition forestry (LCF). The presented variables refer to dominant height (hd), quadratic mean diameter (dg), number of trees (N), carbon (C) and volume (V) of the standing trees before (b) the thinning practice. The products of the thinning operation refer to the volume of the removed trees (V_t) and to the tip debris.

The current silvicultural guidelines with a 4 thinning management regime through the rotation period produce a 25% greater yield, in terms of volume and carbon content, than the yield achieved with the low competition management. In terms of total volume (with the thinning removals included) the difference reduces to just 3.9 m³ ha⁻¹, because the **TYF** scenario presents thinning removals of 135.7 m³ ha⁻¹, while for the **LCF** scenario 212.5 m³ ha⁻¹ are harvested by thinning. These results could lead to an undifferentiated selection between both management options, or even to the preference of the **LCF** scenario as it provides timber of great size (31.8 cm of diameter, for the mean tree, at age 45). Nevertheless, under a

SFM other issues, such as the assessment of risk, need to be taken into consideration. For the example, the ratios of the mean height to mean diameter of the dominant trees vary between 0.57 and 0.60, during the rotation. This indicates a potential problem of stability of the trees under windy conditions and a high risk of wind damages. Therefore, to perform a thinning at age 24, with a density of 1054 trees ha^{-1}, might not be a secure option.

Also shown in Table 1 are the dead wood estimates from downed woody debris, ranging from 9 to 10.1 m^3 ha^{-1}. This is a part of the material obtained by thinning. Other values could be estimated depending on the specification of the merchantability limits and of accounting, for instance, additionally for the mass of the branches. For a complete assessment of the debris produced in the stand, it is suggested to evaluate the debris material using as an indicator the biomass (or carbon content) of the entire crown component (branches and leaves, instead of restricting the evaluation to the tip volume of the trees). Independently of the indicator chosen, the ModisPinaster features allow for the quantification of debris by diameter classes (not shown in Table 1). This might be of importance in some studies, such as when evaluating for the wood decay.

Although dead wood and decaying trees were considered for a long time as being of less or null commercial value, they do have considerable ecological value. The dead wood has a major influence on biodiversity. Many forest species, such as forest floor vertebrates and insects benefit or depend on dead wood material for habitat or resources. The scenario COF is presented as an example of a potential scenario to promote habitat and resources for the conservation of biodiversity. A comparison between the three scenarios is shown in Table 2.

Sim.	t (yrs)	d classes range (cm)	dg (cm)	Total volume (m^3 ha^{-1})	Harvested volume (m^3 ha^{-1})	Deadwood volume (m^3 ha^{-1})
TYF	45	20 - 35	28.2	509.8	135.7	9.0
LCF	45	20 - 35	31.8	505.9	212.5	10.1
COF	45	15 - 35	24.0	582.4	-	183.8
COF	65	25 - 40	31.5	920.1	-	350.7

Table 2. Characteristics of the stand at the end of the rotation age according to the simulation results achieved for the management scenarios of typical forestry guidelines (TYF), low competition forestry (LCF) and of combined objectives forestry (COF). In the TYF and LCF scenarios, the rotation age is fixed at 45 years while in the COF scenario an extended rotation age of 65 years is promoted. The characteristics refer to the diameter distribution (range and quadratic mean diameter, dg) and to the total volume of the standing trees at age t; to the removals obtained by thinning for the management options TYF and LCF; and to the accumulated volume of the trees that die during the simulation period for the three scenarios.

In Table 2, the total volume includes the volume from thinning practices, for the scenarios TYF and LCF, and the volume of the dead trees, for the COF scenario. As expected, a maximum value is achieved when there is no interference in the stand growth. This is consistent with the current consensus about the effect of stand density on growth (Zeide, 2001). The objective of thinning is to anticipate mortality and to provide better growth conditions for the remaining trees. Other goals might be added, such as, to obtain a target value for diameter at the thinning ages and at the final rotation. For the examples presented, the management according to a window for density between 25-35%, presents material distributed by 4 diameter classes (5 cm of amplitude) with average dimension slightly higher than the material obtained with the typical silviculture. When combined objectives are required, other guidelines need to apply. The growth under high densities (55-60% of *SDI*) and the extension of the rotation age, as presented in the **COF** scenario, allow for exploitation of wood, although of minor size, and guarantees better habitat conditions for the promotion of the fauna biodiversity.

4. Conclusion

The use of forest models has undoubtedly enhanced the scientific knowledge about forest dynamics and about the effects of alternative silvicultural options in the stand evolution. Taking as example the ModisPinaster model, it was shown how essential the models are for management decisions and planning purposes. The managers are facing challenges in terms of selecting the most appropriate management guidelines that assure the management goals, which might combine timber and other forest benefits, and increasingly of accounting for risk. Different scenarios are permitted for simulation, leading to better-quality choices under a Sustainable Forest Management guiding principle. From a user's point of view, other needs, such as an easy and free use of the models, are additionally mandatory. Software simulators of forest growth and stand dynamics should favour re-use and share methods and algorithms, promote integration and encourage partnerships. Capsis was delineated to follow these criteria. The examples provided here for ModisPinaster prove how an efficient software simulator can improve capabilities of models and encourage their use by the stakeholders for guidance in decision making.

The involvement in Capsis of different actors, developers, modellers and end-users, brings to the top the desirable features of "easy-to-use" models, allowing for a prompt search of guiding principles while securing the scientific validity of the simulation estimates.

5. Acknowledgment

The first author acknowledges the COST Action FP0603 for the financial support to integrate the ModisPinaster model in the Capsis platform at AMAP, Montpellier. The activities were developed in 2009 and 2011, under the short term scientific missions FP0603_04967 and FO0603_090511-007846-7846, respectively.

Acknowledgments are extended to Prof. M. Tomé for scientific supervision of part of this work, while the first author was preparing her PhD research.

6. References

Alder, D. (1979). A Distance-independent Tree Model for Exotic Conifer Plantations in East Africa. *For. Sci.*, Vol. 25, pp. 59-71.

Alegria, C.M.M. (2003). Estudo da Dinâmica do Crescimento e Produção dos Povoamentos Naturais de Pinheiro Bravo na Região de Castelo Branco. PhD Thesis, Instituto Superior de Agronomia, Universidade Técnica de Lisboa, Lisboa.

Almeida, L.F.R. (1999). Comparação de Metodologias para Estimação de Altura e Volume em Povoamentos de Pinheiro Bravo no Vale do Tâmega. Relatório Final de Estágio, Universidade de Trás-os-Montes e Alto Douro, Vila Real.

Bouffier, L.; Raffin, A.; Rozenberg, P.; Meredieu, C. & Kremer, A. (2009). What are the consequences of growth selection on wood density in the French maritime pine breeding programme? *Tree Genetics & Genomes*, Vol. 5, pp. 11-25.

Bravo, F.; Rodrígues, F. & Ordoñez, A.C. (2010). *SimanFor: Sistema de Apoyo para la Simulación de Alternativas de Manejo Forestal Sostenible*. Retrieved from < www. simanfor.es>.

Brin, A.; Meredieu, C.; Piou, D.; Brustel, H. & Jactel, H. (2008). Changes in quantitative patterns of dead wood in maritime pine plantations over time. *For. Ecol. Manage.*, Vol. 256, pp. 913-921.

Cucchi, V.; Meredieu, C.; Stokes, A.; de Coligny, F.; Suarez, J. & Gardiner, B.A. (2005). Modelling the windthrow risk for simulated forest stands of Maritime pine (*Pinus pinaster* Ait.). *For. Ecol. Manage.*, Vol. 213, pp. 184-196

de Coligny, F. (2007). Efficient Building of Forestry Modelling Software with the Capsis Methodology, *Proceedings of the Second International Symposium on Plant Growth Modelling, Simulation, Visualization and Applications*, pp. 216-222, Beijing, China, November 13-17, 2006.

Decourt, N. & Lemoine, B. (1969). Tables de Production pour le Pin Maritime dans le Su-Ouest de la France. *Revue Forestiére Française*, Vol. 26, No.1, pp. 5-16.

Diéguez-Aranda, U.; Alboreca, A.R.; Castedo-Dorado, F.; González, J.G.A.; Barrio-Anta, M.; Crecente-Campo, F.; González, J.M.G.; Pérez-Cruzado, C.; Soalleiro, R.R.; López-Sánchez, C.A.; Balboa-Murias, M.A.; Varela, J.J.G. & Rodríguez, F.S. (2009). *Herramientas Selvícolas para la Gestión Forestal Sostenible en Galicia*. Consellería do Medio Rural, Xunta de Galicia, Spain.

Dufour-Kowalski, S.; Courbaud, B.; Dreyfus, P.; Meredieu, C. & de Coligny, F. (2012). Capsis: an Open Software Framework and Community for Forest Growth Modelling. *Ann. For. Sci.* (DOI: 10.1007/s13595-011-0140-9).

Echeverría, I. & de Pedro, S. (1948). *El Pinus pinaster en Pontevedra. Su Productividad Normal y Aplicación a la Celulosa Industrial*. Boletines del IFIE, n° 38, Madrid.

Farrell, E.P.; Führer, E.; Ryan, D.; Andersson, F.; Hüttl, R. & Piussi, P. (2000). European Forest Ecosystems: Building the Future on the Legacy of the Past. *For. Ecol. Manage.*, Vol.132, pp. 5-20.

Fonseca, T.F. (2004). Modelação do Crescimento, Mortalidade e Distribuição Diamétrica, do Pinhal Bravo no Vale do Tâmega. PhD Thesis, Universidade de Trás-os-Montes e Alto Douro,Vila Real.

Fonseca, T.F.; Marques, C.P. & Parresol, B.R. (2009). Describing Maritime Pine Diameter Distributions with Johnson's S_B Distribution Using a New All-parameter Recovery Approach. *For. Sci.*, Vol. 55, No. 4, pp. 367-373.

Fonte, C.M.M. (2000). Estimação do Volume Total e Mercantil em *Pinus pinaster* Ait. no Vale do Tâmega. Relatório Final de Estágio, Universidade de Trás-os-Montes e Alto Douro, Vila Real.

Gauquelin, X. & Courbaud, B. (Ed(s).). (2006) *Guide des Sylvicultures de Montagne - Alpes du Nord Françaises*. CemOA Publications, Aubière, France.

Gonçalves, A.C.A. (2003). Modelação de Povoamentos Adultos de Pinheiro Bravo com Regeneração de Folhosas na Serra da Lousã. PhD Thesis, Instituto Superior de Agronomia, Universidade Técnica de Lisboa, Lisboa.

Huang, S.; Morgan, D.; Klappstein, G.; Heidt, J.; Yang, Y. & Greidanus, G. (2001). *GYPSY – A Growth and Yield Projection System for Natural and Regenerated Stands Within an Ecologically Based, Enhanced Forest Management Framework.* Land and Forest Division, Alberta Sustainable Resource Development, Canada.

Johnson, N.L. (1949). Systems of Frequency Curves Generated by Methods of Translation. *Biometrika*, Vol. 36, pp. 149-176.

Lemoine, B. (1991). Growth and Yield of Maritime Pine (*Pinus pinaster* Ait.): the Average Dominant Tree of the Stand. *Annales des Sciences Forestières*, Vol.48, pp. 593-492.

Lopes, D.M.M. (2005). Estimating Net Primary Production in Eucalyptus globulus and Pinus pinaster Ecosystems in Portugal. PhD Thesis, KingstonUniversity, Kingston.

Luis, J.S.; Fonseca, T.F. (2004). The Allometric Model in the Stand Density Management of Pinus pinaster Ait. in Portugal. *Ann. For. Sci.*, Vol. 61, pp. 1-8.

Marques, C.P. (1987). Qualidade das Estações Florestais – Povoamentos de Pinheiro Bravo no Vale do Tâmega. PhD Thesis, Universidade de Trás-os-Montes e Alto Douro,Vila Real.

Marques, C.P. (1991). Evaluating Site Quality of Even-Aged Maritime Pine Stands in Northern Portugal Using Direct and Indirect methods. *For. Ecol. Manage.*, Vol. 41, pp. 193-204.

Meredieu, C.; Dreyfus, P.; Cucchi, V.; Saint-André, L.; Perret, S.; Deleuze, C.; Dhôte, J.F. & de Coligny, F. (2009). Utilisation du Logiciel Capsis pour la Gestion Forestière. *Forêt-Entreprise*, Vol. 186, pp. 32-36.

Meredieu C.; Labbé, T.; Orazio, C.; Bucket, E.; Cucchi, V. & de Coligny, F. (2005). New Functionalities Around an Individual Tree Growth Model for Maritime Pine: Carbon and Nutrient Stock, Windthrow Risk, Log Yield, Wood Quality, and Economical Criteria. Oral presentation for the *IUFRO Working Party S5.01-04 Conference*, New Zealand, November, 2005.

Moreira, A.M. & Fonseca, T.F. (2002). Tabela de Produção para o Pinhal do Vale do Tâmega. *Silva Lusitana*, Vol. 10, No. 1, pp. 63-71.

Najar, M. (1999). Un Nouveau modèle de Croissance pour le Pin Maritime. *Informations – Forêt*, Vol.4, No. 597, pp. 1-6, ISSN 0336-0261.

Nieto, A. & Alexander, K.N.A. (2010). *European Red List of Saproxylic Beetles.* Publications Office of the European Union, Luxembourg. Retrieved from <http://ec.europa.eu/environment/nature/conservation/species/redlist/downloads/European_saproxylic_beetles.pdf>

Orois, S.S. & Soalleiro, R.R. (2002). Modelling the Growth and Management of Mixed Uneven-aged Maritime Pine-Broadleaves Species Forests in Galicia (Northwestern Spain). *Scan. J. For. Res.*, Vol. 17, No. 6, pp. 538-547.

Parresol, B.R. (2003). Recovering Parameters of Johnson's S_B Distribution. *Res. Pap. SRS-31*, USDA, USA.

Parresol, B.R.; Fonseca, T.F. & Marques, C.P. (2010). Numerical Details and SAS Programs for Parameter Recovery of the S_B distribution. *Gen. Tech. Rep. SRS-122*, USDA, USA.

Páscoa, F. (1987). Estrutura, Crescimento e Produção em Povoamentos de Pinheiro Bravo; um Modelo de Simulação. PhD Thesis, Instituto Superior de Agronomia, Universidade Técnica de Lisboa, Lisboa.

Pretzsch H., Grote, R.; Reineking, B.; Rötzer, Th. & Seifert, St. (2008). Models for Forest Ecosystem Management: A European Perspective. *Annals of Botany*, Vol. 101, No. 8, pp. 1065–1087.

Reineke, L.H. (1933). Perfecting a stand-density index for even-aged forests. *J. Agric. Res.*, Vol. 46, pp. 627-638.Santos-Hall, F.A. (1931). *Tabela de Produção Lenhosa para o Pinheiro Bravo*. Separata do Boletim do Ministério de Agricultura, Ano XIII, No.1, 1ª Série, Lisboa.

Salas-Gonzalez, R.; Houllier, F.; B. Lemoine, B. & Pignard, G. (2001). Forecasting wood resources on the basis of national forest inventory data. Application to *Pinus pinaster* Ait. in southwestern France. *Ann. For. Sci.* Vol. 58, pp. 785-802.

Seidl, R.; Fernandes, P.M.; Fonseca, T.F.; Gillet, F.; Jonsson, A.M.; Merganicová, K.; Netherer, S.; Arpaci, A.; Bontemps, J.; Bugmann, H.; González-Olabarria, J.R.; Lasch, P.; Meredieu, C.; Moreira, F.; Schelhaas, M. & Mohren, F. (2011). Modelling Natural Disturbances in Forest Ecosystems: a Review. *Ecological Modelling*, Vol.222, pp. 903-924.

Shevts, V. & B. Zeide, B. (1996). Investigating parameters of growth equations. *Can. J. For. Res.*, Vol. 26, pp. 1980-1990.

Soalleiro, R.R. (1995). Crecimiento y Producción de Masas Regulares de *Pinus pinaster* Ait. en Galicia. Alternativas Selvícolas Posibles. PhD Thesis, Universidad Politécnica de Madrid, ETS de Engenieros de Montes, Madrid.

Soalleiro, R.R.; González, J.G.A. & Vega, G. (1994). *Piñeiro do País: Modelo Dinâmico de Crecemento de Masas Regulares de Pinus pinaster Aiton en Galicia (Guía para o Usuario do Programa PINASTER)*. Capacitación e Extensión. Serie Manuais Prácticos 8. Consellería de Agricultura, Gandería e Montes, Xunta de Galicia, Santiago de Compostela.

Zeide, B. (2001). Thinning and Growth: a Full Turnaround. *Journal of Forestry*, Vol. 99, No. 1, pp. 20-25.

Zeide, B. (2008). The Science of Forestry. *Journal of Sustainable Forestry*, Vol. 27, No. 4, pp. 345-473.

7

Individual-Based Models and Scaling Methods for Ecological Forestry: Implications of Tree Phenotypic Plasticity

Nikolay Strigul
Stevens Institute of Technology
USA

1. Introduction

The concept of sustainable forest management (SFM) has been developed across traditional disciplinary boundaries, including natural resource management, environmental, social, political, economical, climatic sciences and ecology. The Montreal process (www.mpci.org) has established multidisciplinary criteria for the SFM of temperate and boreal forests. In parallel with the Montreal process, the pan-European forest policy process (www.foresteurope.org, Forest Europe, The Ministerial Conference on the Protection of Forests in Europe, MCPFE) has developed criteria for SFM in Europe. Practical implementation of SFM criteria requires the development of scaling methods to link individual-level processes, pollution effects, climatic changes and silvicultural operations to large-scale ecosystem patterns and processes. A general problem is that data obtained in numerous experimental studies that address effects at the individual level cannot be translated to the ecosystem level without a large amount of uncertainty. Forested ecosystems have a complicated spatially heterogeneous hierarchical structure emerging from numerous interdependent individual processes. The fundamental ecological questions are how macroscopic patterns emerge as a result of self-organization of individuals and how ecosystems respond to different types of environmental disturbances occurring at different scales (Levin, 1999).

The SFM employs the ecological forestry (EF) silvicultural approach, which is significantly distinct from the intensive (traditional) forestry and, therefore, requires different modeling tools than traditional forestry models. Traditional or intensive forestry is focused on wood production to maximize productivity of land use and usually involves tree plantations of commercially important trees (Nyland, 1996; Perry, 1998). Different silvicultural tools help increase wood fiber production. In particular, use is made of fast growing and disease resistant cultivars, vegetation control via thinning and regeneration harvesting techniques, soil management, and forest pests and noncrop vegetation control. Intermediate cutting operations include low, crown and mechanical thinning target future stand growth on higher valued trees to improve the stand yield at final harvest while providing some financial return on the shorter time scales. Traditional forestry also employs prescribed fire, cutting and application of herbicides for regulation of species composition and promoting growth of economically important tree species in the mixed stands.

This chapter is focused on modeling tools for the SFM and EF. The objective of this approach is the optimization of land use (such as wood production and carbon storage) while maintaining biocomplexity of forested ecosystems. The models discussed in this chapter are to be implemented within the SFM framework to optimize land use (such as wood production and carbon storage with the criteria 2 and 5 of Montreal process "Maintenance of productive capacity of forest ecosystems" and "Maintenance of forest contribution to global carbon cycles", respectively; and criteria 1 and 3 of the MCPFE process (Ministerial Conference on the Protection of Forests in Europe) "Maintenance and appropriate enhancement of forest resources and their contribution to global carbon cycles" and "Maintenance and encouragement of productive functions of forests", respectively. The fundamental challenge for ecological forestry is to effectively manage a forest - complex ecological system, rather than a plantation of trees as in the traditional forestry approach. The biocomplexity challenges for ecological forestry are the understanding of why different plant species coexist, and which forces drive forest community structure and dynamics. One of the keystones of ecological forestry is the development of forest management systems in concert with natural processes in forested ecosystems, such as natural disturbances, forest dynamics and succession (Franklin et al., 2007). In particular, development of regeneration harvest approaches that have ecological effects similar to natural disturbances has been considered crucial for ecological forestry. Natural disturbances may occur at different spatial scales resulting in heterogeneity of forested ecosystems. The most common natural disturbances include wind-related disturbances on the individual (forest gaps) and large-scale (for example created by hurricanes and tornadoes), fire-related disturbances, and pest or disease related disturbances. These disturbances may significantly alter ecosystem structure and dynamics; however even the most dramatic events do not completely destroy ecosystems. Certain biological patterns or biological legacies, specific for each type of disturbance, remain unchanged and facilitate forest post-disturbance recovery.

Forest heterogeneity, which emerges as the result of various disturbances, is an essential element of ecological forestry, in contrast to the traditional approach, where stands are spatially homogeneous to reduce tree competition and improve timber quality (Oliver & Larson, 1996). Morphological plasticity allows trees to compete with neighbors and survive in a heterogeneous environment. In particular, open-growing trees, as well as trees growing in plantations without intense crown competition, tend to have symmetrical crowns, straight trunks, and, as a result, high quality timber. Trees growing in mixed spatially heterogeneous stand tend to exhibit plasticity patterns as every individual tree needs to adjust to its local unique neighborhood. These trees have much less value in term of timber than plantation trees. Such trees often have non-symmetrical crowns and curved trunks as they lean towards sunlight due to intense individual tree competition.

Forested ecosystems demonstrate multiple-scale self-organization patterns in response to disturbances. At present, we lack the predictive modeling tools that can combine the effects of forest disturbances occurring at different scales. An ideal model would present an analytically tractable model predicting landscape-level vegetation dynamics using individual ecophysiological traits as variables and available forest survey data as initial conditions. How can we develop such models?

Simon Levin, in a seminal paper (Levin, 2003), considered a modern theoretical approach to multiscale ecological modeling. In particular, he introduced ecological systems as complex adaptive systems which result from self-organization on multiple levels, where individual organisms are linked through interactions between each other and the abiotic environment. In this chapter the framework of complex adaptive systems is applied to forest ecology and management. The forest is considered as a complex adaptive system (as a mosaic of individual plants, each of which grows adaptively in its biotic and abiotic environment in dynamic interaction with its neighbors). These interactions occur simultaneously at different temporal and spatial scales, both above and below ground, and lead to the development of self-organized patterns and structural complexity. The central question of this chapter is how forest patterns emerge as a result of the self-organization of individual trees. Individual tree plasticity is a critical process for forest modeling, though it has previously not been taken into account. The plasticity patterns of tree crowns in response to light competition enable directional growth toward available light, and lead to tree asymmetries caused by stem inclinations and inhomogeneous branch growth. Recently developed individual-based forest simulators, Crown Plastic SORTIE (Strigul et al., 2008) and LES, focus on forest self-organization at the stand level. These models, by incorporating individual crown plasticity, predict substantially different macroscopic patterns than do previous models (regularity in canopy spatial structure, for instance, which has only recently been noticed in field studies). Most importantly, the simulator's structure allows to derive an accurate approximation of the individual-based model, the Perfect Plasticity Approximation (PPA). This macroscopic system of equations predicts the large-scale behavior of the individual-based forest simulator, using the same parameter values and functional forms (Strigul et al., 2008). In particular, the PPA offers good predictions for 1) stand-level attributes, such as basal area, tree density, and size distributions; 2) biomass dynamics and self-thinning; and 3) ecological patterns, such as succession, invasion, and coexistence.

This chapter also introduces a theoretical framework for the scaling of forest spatial dynamics from individual to the landscape level based on the PPA model. The major objective of this approach is to scale up forest heterogeneity patterns across the forest hierarchy. The major idea is that the forest dynamics at the landscape level can be modeled by separating dynamics within forest stands caused by individual-level disturbances from the dynamics of the stand dynamics caused by large disturbances. The model, called Matreshka (after the Russian nesting doll) employs the PPA model as an intermediate step of scaling from the individual level to the forest stand level (or patch level). To describe the patch dynamics at the next hierarchal level, i.e., the forest stand mosaic, we employ the patch-mosaic modeling framework (Strigul et al., 2012). The Markov chain model for the mosaic of forest stands in the Lake states (MI, Wi, and MN) has been recently parameterized using the FIA data (Strigul et al., 2012). The Matreshka model unites already known models and uses the notion of ecological hierarchy that has been widely employed in landscape ecology (Bragg et al., 2004; Clark, 1991; Wu & Loucks, 1996).

The chapter consists of three sections: Section 2 introduces tree morphological plasticity as a fundamental pattern for the canopy self-organization. In section 3 the individual-based modeling approach is considered with a special focus on the development of the Crown Plastic SORTIE model, LES, and PPA models. Section 4 introduces a theoretical framework for the scaling of forest dynamics from individual to the landscape level based on the PPA model.

2. Individual tree plasticity and canopy self-organization

2.1 Crown plasticity and leaning of individual trees

In competing for light, trees invest carbon and other resources to achieve such above-ground form as will provide them with enough light for photosynthesis. To achieve this goal, individual trees demonstrate amazing phenotypic plasticity, using advantages of modular organization (Ford, 1992). Numerous factors constrain tree development, for instance, gravity

Fig. 1. Tree growing across a small forest river. A typical example of tree plasticity and "riverside behavior". The Institute for Advanced Study Woods (Princeton, NJ).

(McMahon & Kronauer, 1976) and other abiotic factors such as wind (Grace, 1977) and snow (King & Loucks, 1978), neighborhood effects (Ford, 1992), and, also genetic and physiological constraints such as the need to provide an efficient connection between their own above- and below-ground parts (Kleunen & Fischer, 2005). One permanent goal of a given individual tree is to develop an optimal crown under the current limitations in the dynamic environment. This includes different wood-allocation strategies in open-growing trees and trees in dense stands (Holbrook & Putz, 1989), as well as the development of sun branches and the degradation or physiological modification of shaded branches (Stoll & Schmid, 1998).

The physiological mechanisms underlying plant-phenotypic plasticity and phototropism have received significant attention in recent decades, yet many phenomena remain unclear (Firn, 1988; Kleunen & Fischer, 2005). Apical control can partially explain interspecific differences in tree leaning, crown shapes, and also differences in growth patterns between understory and overstory trees (Loehle, 1986; Oliver & Larson, 1996). Plants have also a variety of photosensory systems to detect their neighbors and select an optimal growing strategy. A better-investigated, phytochrome-signaling mechanism triggers some adaptive morphological changes such as adaptive branching (Stoll & Schmid, 1998) and stem elongation (Ballaré, 1999) in response to the alterations in far-red radiation caused by the reflection of sunlight by neighbor plants.

Tree-plasticity patterns relating to the competition for light and phototropism include the development of an asymmetrical crown, as a result of both the growth of individual branches and the phototropism of the whole tree (resulting in trunk elongation and inclinations). Tree-plasticity patterns caused by competition for light are more pronounced near forest margins, such as road cuts or riverbanks (Fig. 1). At these places trees develop asymmetric crowns and lean toward the gap, this pattern was called "riverside behavior" (Loehle, 1986).

Crown asymmetries and tree leaning can be also caused by factors not related to light competition, for example, soil creep (Harker, 1996), wind (Lawrence, 1939), and destruction of the apical meristem by insects. Trees growing on hillsides often have special trunk inclinations induced by soil creep, which geologists call a "d" curve (Harker, 1996). In this case, the base of the tree trunk starts at an angle to the vertical, with this angle continuously decreasing toward the top of the tree. However, such trees can have symmetrical crowns. This type of curved trunk is used as an indicator of soil creep. It was suggested that downward trunk inclinations of understory trees growing on the slope may have adaptive significance for light competition (Ishii & Higashi, 1997), however other authors disagree with this hypothesis (Loehle, 1997). While crown asymmetry and trunk inclination represent two closely related patterns providing for tree morphological plasticity in light competition, the development of asymmetrical crown has received more attention than the tree leaning process. It was recognized since the earliest stages of the forest science development that an understanding of how tree crown is changed in competition for light is critical for forest growth predictions (Busgen & Munch, 1929; Reventlow, 1960). Tree crown area is naturally connected with total leaf surface, photosynthetic activity, carbon gain, and tree growth (Assmann, 1970; Smith et al., 1997). Crown competition is analyzed using crown class classification, individual tree zone of influence and by computing different competition indices. In forestry practice these methods are applied under the implicit assumption that trees grow vertically and the center of the zone of influence is also the center of tree growth. This is an important assumption in silviculture, since traditional foresters typically considered curved-trunk trees to be abnormal and unconditioned, and ignored them (Macdonald & Hubert, 2002; Westing & Schulz, 1965). Methods to measure such trees were also not developed (Grosenbaugh, 1981). The main objective of traditional silviculture to produce qualitative wood from well-formed trees (i.e., trees with symmetrical crowns and straight stems). Therefore for foresters tree leaning is in fact a problem which causes the development of bad-formed trees, rather than an important ecological property (Macdonald & Hubert, 2002). Typical planting and thinning regimes in silviculture significantly reduce the frequency of trunk inclinations (Assmann, 1970; Oliver & Larson, 1996; Smith et al., 1997). Only a few studies are concerned with adaptive trunk inclinations associated with the phototropism of the whole tree. In the beginning of the last century these patterns were described by German forester Arnold Engler (Engler, 1924). More recently, Loehle (Loehle, 1986) reported connections between trunk inclinations and the phototropism of the whole tree, based on data collected in Georgia and Washington State.

2.2 Gap dynamics and community-level patterns

Forest gaps are defined as small, localized disturbances, such as treefalls, which cause asynchronous local-forest regeneration processes (Oliver & Larson, 1996). In contrast, large-scale catastrophic disturbances, such as hurricanes or clearcutting, cause synchronized forest regenerations on the stand level. Gap dynamics is an important ecological process in which tree-plasticity patterns are exhibited (Fig. 2). Since forest gap dynamics constitutes a major process of regeneration, succession, and species coexistence (McCarthy, 2001; Ryel & Beyschlag, 2000), tree plasticity patterns can be associated with major trade-offs determining the strategies of trees.

Typically, large trees located at the gap border extend their crowns toward the gap (Hibbs, 1982) significantly reducing gap size and affecting canopy recruitment (Frelich & Martin, 1988). Gap closure, in turn, involves an interplay between two processes. The first process consists of lateral gap closure, brought about by crown encroachments of large trees at the

Fig. 2. A typical forest gap in the Institute for Advanced Study Woods (Princeton, NJ). Trees growing on the gap boundaries demonstrate plasticity patterns and phototropism, they modify their crowns and lean toward the gap.

gap borders and growth; the second process involves the crown development of small trees in the gap. The relative contributions of these two processes can be regulated by the gap size and species composition of both saplings and neighbor trees. To capture a small gap, saplings must be able to grow fast enough to compete with expanding crowns of dominant and co-dominant trees at the gap borders; in large gaps, by contrast, saplings have more opportunities to establish a canopy (Cole & Lorimer, 2005; Gysel, 1951; Webster & Lorimer, 2005; Woods & Shanks, 1959).

Individual tree plasticity leads to the development of a regular spatial canopy structure; in particular, crown centers are spaced more evenly than are the bases of plants. This pattern was reported for natural forest stands on Hokkaido (Ishizuka, 1984), and in the pure stand of *Atherosperma moschatum Labillardiere* (Monimiaceae) in Tasmania (Olesen, 2001), where crown-center distributions of all canopy were close to the uniform. Similar patterns were also discovered in the natural mature *Pinus sylvestris L.* forest in Eastern Finland (Rouvinen & Kuuluvainen, 1997); however, in that case the direction of crown asymmetry was strongly weighted in a southern and southwestern direction, which is the direction of most abundant solar radiation. It was suggested that in this forest, both factors, i.e., light competition with neighbors and phototropism toward the south, led to crown asymmetry and regular crown spacing patterns (Rouvinen & Kuuluvainen, 1997). Regular spacing of crowns in the canopy has been also established in computer simulations, where individual plants are able to exhibit adaptive crown plasticity (Strigul et al., 2008; Umeki, 1995a). However, forest simulators that does not include tree plasticity do not predict canopy regularity (Strigul et al., 2008).

2.3 Interspecific differences and cost of tree plasticity

Most tree species of different systematic and ecological groups demonstrate some tree plasticity patterns. It has been reported, for instance, for conifers (Loehle, 1986; Stoll & Schmid, 1998; Umeki, 1995b) and broad-leaf trees (Brisson, 2001; Woods & Shanks, 1959), in tropical (Young & Hubbell, 1991) and temperate forested ecosystems (Frelich & Martin, 1988; Gysel, 1951; Stoll & Schmid, 1998; Webster & Lorimer, 2005). At the same time, different tree species vary significantly in their ability to execute plasticity patterns; this raises questions concerning the different life histories and ecological strategies associated with tree plasticity and light competition. In particular, gap closure by crown encroachment of adjacent dominant and co-dominant trees was reported to be a typical process in the replacement of chestnut (*Castanea dentate* (Marsh.) Borkh.) by *Quercus prinus* L. and *Q. rubra* L. in the Great Smoky Mountains (Woods & Shanks, 1959). Northern red oak *Q. rubra* L. significantly surpassed yellow poplar (*Liriodendron tulipifera* L.) in its capacity for crown encroachment (lateral extension rates are 16.5 cm/year and 9.2 cm/year respectively) in Appalachian hardwood stands (Trimble & Tryon, 1966). The average lateral crown growth toward the small tree gaps of seven tree species in hemlock-hardwood forests in Massachusetts varied from 6 to 14 cm/year (Hibbs, 1982). *Quercus rubra* L. demonstrated the fastest lateral crown growth, with an average 14.03 ± 1.65 cm/year and a maximum 26.4 cm/year. The other six species were ranked according to their average lateral crown growth (in cm/year) toward the gap, as follows: *Betula papyrifera* Marsh. $10.87 \pm 1.39 > B. lenta$ L. and *B. alleghaniensis* Britt. $10.68 \pm 1.58 > Tsuga\ Canadensis$ (L.) Carr. $10.68 \pm 1.58 > Acer\ rubrum$ L. $8 \pm 0.72 > Pinus\ strobus$ L. 6.10 ± 0.94. Average annual crown lateral extensions toward the gaps of 13 tree species in the Southern Appalachians (Runkle & Yetter, 1987), varied from 8.6 cm/year (*Fraxinus americana* L.) and 13.1 cm/year (*Tsuga Canadensis* (L.) Carr.) to 31.4 cm/year (*Magnolia fraseri* Walt.) and 28.7 cm/year (*Acer rubrum* L.). Three species had shown a lateral extension rate of more than 20 cm/year (*B. alleghaniensis* Britt. 22.3 cm/year, *Liriodendron tulipifera* L. 21.8 cm/year, and *Acer saccharum* 20.8 cm/year Marsh.), and the other six broad-leaved tree species showed very similar rates of $17.1 - 18.8$ cm/year (Runkle & Yetter, 1987). This brief review demonstrates that lateral crown growth rate toward the gap can vary between the stand and tree species. While some species (for example, *Q. rubra* L.) typically demonstrate more plasticity than others, many species exhibit similar patterns, and some (for example, *T. Canadensis* (L.) Carr.) apparently have much less ability to extend their crowns toward the gap. These estimates are employed in the LES model (section 3.2)

German foresters' studies of the first half of the 20th century (see Engler (1924), Busgen & Munch (1929) p. 41, Assmann (1970) pp. 244, 284, 348 and subsequent references) found that conifer trees are less plastic than broad-leaved trees, which are capable of filling highly variable types of growing space. To account for these differences, it was suggested that broad-leaved trees, such as oaks and beeches, exhibit more phototropism than conifers, such as spruces and silver firs, which have "extremely energetic geotropism" (Assmann, 1970; Busgen & Munch, 1929). This conclusion is supported by the later studies (Loehle, 1986; Umeki, 1995b). It was suggested that the contrast plasticity patterns of conifers and broad-leave trees can be explained by the apical control differences (Loehle, 1986; Waller, 1986).

One important open problem is the lack of quantitative estimates of physiological traits associated with tree plasticity. Gravity is the universal force affecting tree form and growth. This force favors a vertical trunk and a symmetrical crown, which compose the typical form for an open growing tree. In this case the crown center of mass and the tree base are located on the same vertical line, which is the axis of tree symmetry. The execution of tree plasticity patterns, such as adaptive growth of branches and tree leaning, results in tree asymmetries and changes of the crown mass center that can make the tree less stable. Then, crown asymmetries and tree leaning should have some additional cost per tree compared to a symmetrical crown expansion (Busgen & Munch, 1929; Olesen, 2001). In a wet lowland tropical forest, tree asymmetry can increase the likelihood of the tree fall (Young & Hubbell, 1991). Tree anatomy studies and mechanical considerations show that the development of tree asymmetry causes stem tensions which should correlate with the development of additional structural tissues (Ford, 1992; McMahon & Kronauer, 1976). Umeki (Umeki, 1995a) assumed that the cost of tree asymmetry can be expressed by a reduction of tree height proportionally to the distance of the crown center movement. This assumption is also made in the Crown Plastic SORTIE and LES models (section 3.2). Loehle (Loehle, 1997) assumed that small trees with an elastic trunk can grow at an angle at practically no cost, and suggested that cost estimations are important only for large trees.

3. Scaling of vegetation dynamics: from individual trees to forest stands

The mainstream research approach in modern forestry is to use mathematical modeling in concert with experimental approaches. Certain limitations of experimental approaches make mathematical modeling especially useful. In particular, in experimental studies it is often necessary to concentrate on one focal level of organization while ignoring processes at other scales. Conclusive experimental results to support land-management decisions on different silvicultural techniques may not be obtained on a reasonable time scale and can be too expensive. Despite the availability of different forest models for use in either traditional forestry or in ecological studies, these models are often not suitable for ecological forestry. Forest yield tables is one of the oldest biological models with more than a 200-year history of development and practical applications to plantations with reduced tree competition (Mitchell, 1975; Shugart, 1984). However, forest yield tables is of an empirical nature and limited applications to more spatially heterogeneous silvicultural systems with intensive crown competition. Individual-based models (IBMs) simulating stand development emerged in the 1960s, when computer technology allowed for doing spatially-explicit simulations. Spatially explicit models can incorporate processes that occur at different scales and predict the dynamics of a forest by predicting each individual's birth, dispersal, reproduction and death and how these events are affected by spatial competition for resources with neighbors. Forest growth IBMs were developed in different directions. Foresters have developed stand simulators in order to estimate and optimize stand production; meanwhile, ecologists needed tools to study succession, species coexistence, and dynamics of indigenous forests. This difference in initial goals is reflected in the model structures, as forester and ecological models each concentrate on different aspects of forest development. Ecological models, such as the family of gap models originated from JABOVA include detailed descriptions of ecological processes which are considered to be most important, such as succession and gap dynamics (Botkin, 1993; Shugart, 1984). Forester IBMs, such as TASS (Mitchell, 1975), focus on overstory dynamics and on detailed descriptions of individual tree growth in the given neighborhood, which is important for plantations, ignoring seed production, gap and understory dynamics.

3.1 Individual-based forest simulators and tree plasticity

With respect to crown competition, individual-based forest simulators embody a wide range of assumptions. JABOVA-FORET models and many of their descendants, such as gap models, are based on the premise that a forest can be represented as a mosaic of homogeneous patches, i.e., gaps, each of which can be modeled independently. The size of every gap is usually assumed to be equal to the size of one large overstory tree. These patches have a horizontally homogeneous structure-i.e., the crowns of all trees in a gap extend horizontally over each patch (Botkin, 1993; Bugmann, 2001). The SORTIE model, descended from the JABOVA-FORET family, is a gap model in which trees in the gap have explicit spatial crowns (Pacala et al., 1996). The aboveground part of a single tree in SORTIE is represented as a rigid cylindrical crown, described by a species-specific radius and a crown depth around the vertical trunk, tree-plasticity patterns are not included (Pacala et al., 1996). This representation allows for the simulation of both light distribution in the canopy and tree growth in accordance with the availability of light, depending on local light heterogeneity. Numerous individual-based stand simulators employ the zone of influence concept (Biging & Dobbertin, 1995; Bugmann, 2001; Mitchell, 1980), and crown competition is often accounted for by means of calculation of competition indices (Burton, 1993; Liu & Ashton, 1995). A zone of influence is usually defined as a circle around a tree center, where a focal tree can interact with its neighbors. This concept was used in studies of above-ground and below-ground competition (Aaltonen, 1926; Biging & Dobbertin, 1995; Casper et al., 2003). In the 19th century the term "crown ratio" was introduced to describe the ratio between d.b.h. and the average crown spread of a tree (Lane-Poole, 1936). This parameter was used as a stand characteristic reflecting the intensity of light competition in every crown class to optimize the thinning strategy, by reducing crown competition in silviculture practice (Krajicek et al., 1961; Lane-Poole, 1936). Later, the dominant-tree class was replaced by open-grown trees as the universal standard of trees which are not affected by their neighbors (Krajicek et al., 1961), and the crown area of open-grown trees was defined as a zone of influence for all trees with similar d.b.h. (Biging & Dobbertin, 1995). Comparison of zone of influences with realized dimensions yields different quantitative characteristics, the so-called "crown competition indices" (Biging & Dobbertin, 1995; Krajicek et al., 1961). Individual competition indices, calculated for a representative sample of trees from every crown class, can be averaged to produce a competition measure at the stand level. This scaling approach has several inherent limitations due to the static nature of competition indices, which restricts their usefulness in both practical silviculture and forest ecology (Burton, 1993).

The forest simulators employing the competition indices were united in a class of tree-stand models; in contrast to the crown-stand models (Mitchell, 1980), this old classification emphasizes the importance of simulating the crown and bole development. The crown-stand simulator TASS (Mitchell, 1969; 1975) employs the crown and bole as primary operating units. Crown competition in TASS is calculated as a result of the spatial intersection of the crown-profile functions of neighborhood trees. Similar crown competition algorithm was independently developed for modeling of *Eucalyptus obliqua* stands (Curtin, 1970). This modeling approach was employed in the Crown Plastic SORTIE (Strigul et al., 2008).

The next step in enhancing the realism of crown-plasticity representation is to explicitly simulate the growth of individual branches, instead of calculating a generalized crown-profile function. In 1980, K.J. Mitchell (Mitchell, 1980) included branch-stand models in the stand-model classification; however, such models were not yet developed, due to unrealistic computational resource demand. Technological progress made such models possible, and recent branch-level models were widely used in simulations of the development of form of individual plants and the simplest, evenly distributed, even-aged single-species stands (Godin, 2000; Takenaka, 1994). Several models have been developed to simulate the effects of crown plasticity caused by independent-branch development at the stand level. The WHORL model simulates a two-dimensional forest, where an open-tree crown is represented a system of horizontal disks, simulating a crown layer (Ford, 1992). Disks and their sectors can grow and die independently depending on local light availability in the stand. As a result, a tree crown in the stand develops as an asymmetrical system of whorls stacked along a central vertical axis, representing the tree trunk. A similar crown representation, using a pyramid of independently growing discs (which are also represented by independently growing segments), was employed in the BALANCE model (Grote & Pretzsch, 2002). The LES model (section 3.2) belongs to this group of models; as the next-generation model, it simulates indigenous forests with multiple species (typical simulations are 1000 years of 1 ha plots).

Stand models such as TASS, WHORL, and BALANCE as well as SORTIE and other gap models share a similar assumption concerning tree growth: In these models, trees are assumed to grow vertically, and the zone of influence is centered at the stem base. As a result, these models do not allow for tree leaning as a mechanism of adaptive tree-morphological plasticity. An alternative approach to simulate crown plasticity was developed by K. Umeki, using the crown-vector notion proposed by S. Takiguchi (see Umeki (1995a) for details and cross references). The crown vector is the vector between the stem base position and the centroid of the projected crown area of an individual tree. The centroid's coordinates were calculated using a competition index based on a circular zone of influence (Umeki, 1995a). This approach is also employed in the Crown Plastic SORTIE and LES models to simulate changes of crown center of mass (Fig. 4) .

This brief review demonstrates that a number of individual-based forest simulators vary in their attention to the tree-morphological plasticity patterns. Ecological models, such as SORTIE, describe tree growth in great detail as it relates to fine-scale resource heterogeneity and competition, seed production, and dispersion; however, they ignore both crown competition and tree plasticity. Crown-stand simulators such as TASS provide a detailed description of crown competition and of the underlying-branch plasticity patterns; they do not include ecological patterns, however, and they ignore tree leaning. Finally, the crown-vector approach (Umeki, 1995a) represents a simple and convenient method for simulating tree-leaning patterns.

The Crown Plastic SORTIE model (Strigul et al., 2008) combines the advantages of ecological and forest management IBMs considered above, and incorporates tree plasticity patterns. In particular it includes all the ecological complexity from the SORTIE model, a crown competition algorithm similar to the TASS model, and a crown plasticity algorithm based on the crown-vector approach. This IBM is suitable for predicting prescriptions of ecological forestry concerning management of multi-species and multi-age stands. This IBM gives more realistic predictions than the previous models; in particular, it allows for the observation of canopy regularity patterns emerging as a result of canopy self-organization (Strigul et al., 2008). At the same time, in more simplified simulations without crown plasticity algorithm, crown plastic SORTIE gives the same predictions as SORTIE or TASS depending on the model parameterization employed. This model also allowed derivation of tractable macroscopic equations for forest growth called the Perfect Plasticity Approximation (Strigul et al., 2008). The next generation individual-based model, LES, is introduced below.

3.2 The LES model

An individual-based forest simulator called LES (after the Russian word for forest) simulates spatially explicit tree competition above ground for light and below ground for water and nutrients. The LES model is based on the crown plastic SORTIE model, but operates at the individual branch and root levels (Fig. 3). Trees in the LES model execute phenotypic plasticity patterns considered in section 2. In this model, trees adaptively develop their crowns and root systems to their own unique local neighborhoods.

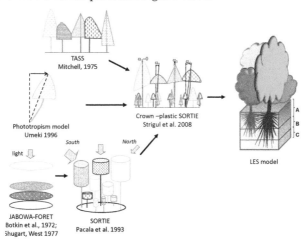

Fig. 3. Genealogy of the LES model

The most important new elements of the LES model compared to its predecessors (Fig. 3) are the following: 1) An individual tree develops a unique crown and root system within a local neighborhood to optimize spatial resource acquisition and allocation. 2) Vertical forest stratification emerges from the branch level competition. The model simulates the development of canopy, midstory and understory levels, allowing for tree classification as dominant, codominant, intermediate and suppressed trees. 3) Tree root systems are described by individual roots, and a vertical soil stratification emerges from individual root competition in three distinct soil horizons.

With respect to crown competition the Crown Plastic SORTIE model (Strigul et al., 2008) includes two essential elements: 1) Crown parametrization and competition algorithm similar to the TASS model, and 2) phototropism algorithm similar to one developed by Umeki (Umeki, 1995a). This model assumes that every tree has a species-specific potential crown shape, which is rotation-symmetrical about the vertical axis through the center of the crown. The realized tree crown is part of the potential crown determined by the spatial tessellation algorithm (Strigul et al., 2008). The advantage of this crown representation is that it leads to the computationally simple and fast algorithm as a horizontal cross section of the potential crown at any height is a circle. However, the major disadvantage is that the Crown Plastic SORTIE assumes the existence of a symmetrical potential crown shape for any tree growing within the forest stand. In the LES model this assumption is relaxed, and the individual crown shape develops as the result of adaptive tree growth within the unique local neighborhood. The new crown algorithm introduced in the LES model (Fig. 4) results in the development of a more realistic canopy than in the Crown Plastic SORTIE model.

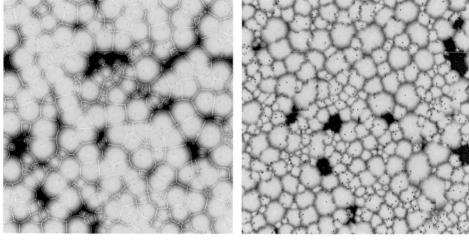

(a) Canopy simulated with the Crown Plastic   (b) Canopy simulated with the LES model.
SORTIE model.

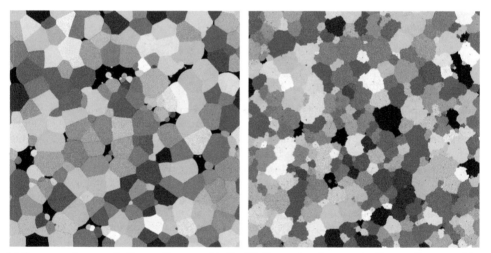

(c) the crown shape projection on the ground in the   (d) the crown shape projection on the ground in
Crown Plastic SORTIE simulations.   LES model simulations.

Fig. 4. Forest canopy simulations in the Crown Plastic SORTIE model (Strigul et al., 2008) and the LES model. In the Crown Plastic SORTIE model (a and c) a realized tree crown is determined as a part of the symmetrical potential crown by the tessellation algorithm (Strigul et al., 2008). In the LES model an individual tree crown develops as the result of adaptive spatial crown development within the unique local neighborhood. Tree crown develops on three hierarchical levels: leave and small branches, large branches (represented as independent spatial sectors) and crown level, represented by the crown center of mass. The figure demonstrates two different canopy visualizations: height-density plots (a and b) and crown ground projection plots (c and d) of canopy trees in a simulated White Pine forest stand (0.25 ha) 200 years after a major disturbance. The brightness level in figures a and b indicates the crown height at every point.

The LES model simulates a tree crown as a hierarchical three-dimensional spatial structure that develops and changes in response to environmental conditions on multiple levels. The LES model incorporates adaptive and random changes on three structural levels: 1) small branch and leaf level, 2) large branch level, 3) crown level. The first level of the crown organization in the LES model corresponds to leaves and small branch level, where every point represents an area of approximately 10 cm^2. Every such crown unit is represented as a point on a two-dimensional grid with the height of this crown component as a parameter. The canopy competition occurs independently at every point, where the highest crown wins the spatial competition. The second level of crown organization in the LES model is the level of large branches represented as independent crown sectors. Every sector is characterized by its width, the height of the lowest leaves and the leave distribution profile within the sector. These characteristics of every individual crown sector are determined by the results of spatial competition in the given neighborhood. The model can simulate crowns with 2^n sectors; most of the simulations are conducted with 8 crown sectors. The largest level of crown organization is the crown level determined by the center of the crown (center of mass); its position is determined by the algorithm of phototropism and crown leaning developed in the Crown Plastic SORTIE model (Strigul et al., 2008).

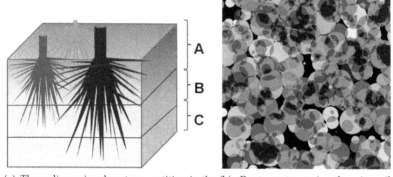

(a) Three dimensional root competition in the LES model.

(b) Root systems in the A soil horizon. 40 years after a major disturbance.

Fig. 5. Simulation of underground root competition in the LES model. A tree root system develops in three soil horizons: top (A), intermediate (B) and low (C). The soil within each horizon is represented as a collection of disjoint spatial units (cuboids), where each soil unit has its own available water content and can be occupied by roots of one or several trees competing for water and nutrients. An individual root system develops independently in different spatial directions corresponding to large roots (simulated as spatial sectors). Trees optimize water uptake by investing available resources in growth of the most efficient root sectors in different soil horizons.

The Crown Plastic SORTIE and all its ancestors focused entirely on the tree competition for light, and ignored below-ground competition for water and nutrients. In the LES model trees have spatial three-dimensional root systems and compete for water and nutrients (Fig. 5(a), 5(b)). Therefore tree growth and resource allocation can be simulated depending on multiple resource limitations (Fig. 6), and, in particular, carbon and water balance are considered at the tree level. The major patterns of belowground tree competition in the LES model are: 1)

The three independent soil horizons (A, B and C on Fig. 5(a)), 2) The spatially heterogeneous water/nutrient distribution within horizons, 3) Several trees can occupy every unit of soil, 4) Directional root growth within each horizon, 5) Individual trees optimize root system growth in three dimensions according to competition constraints and resource availability.

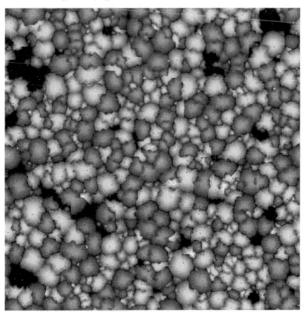

Fig. 6. Canopy level in the LES model simulation of stand development over 200 years, where trees compete for light and water simultaneously. Two different tree species that are colored grey and brown when trees are water-limited, and green and yellow when trees are light-limited, respectively. Most of the canopy trees are water-limited; only several trees with insufficient crowns are light-limited.

3.3 The Perfect Plasticity Approximation (PPA) model
Forest simulation models are effective tools in scaling individual-level spatio-temporal processes to the stand level because they are able to simultaneously incorporate tree ecophysiological traits such as carbon allocation, and capture tree level disturbances and gap dynamics. Individual-based models can also be applied to simulate vegetation dynamics at the landscape level using GIS-based inputs. The major disadvantage of forest individual-based models is that these spatial stochastic processes are not analytically tractable, so their general properties and sensitivities to the choice of parameters and functional forms are uncertain. However, analytically tractable approximations of individual based forest simulators can be developed. In particular, the Perfect Plasticity Approximation (PPA, Strigul et al. (2008)) is a recently developed model predicting the stand-level forest dynamics by scaling up individual-level processes. The PPA offers good predictions for 1) stand-level attributes, such as basal area, tree density, and size distributions; 2) biomass dynamics and self-thinning; and 3) ecological patterns, such as succession, invasion, and coexistence. The model includes a system of von Foerster partial differential equations and the PPA equation.

Unlike the individual-based simulator, the PPA model is both analytically tractable and computationally simple. Initially the model was developed as an approximation of the crown plastic SORTIE model (Strigul et al., 2008), but it was also demonstrated that the PPA model captures the dynamics of the temporary forests. Purves at al. (Purves et al., 2008) estimated the parameters of the PPA model by using the data collected by the Forest Inventory and Analysis (FIA) Program of the U.S. Forest Service (FIA data) for the US Lake states (Michigan, Wisconsin, and Minnesota). It was demonstrated that the PPA model, applied even in its simplest form, carefully predicts forest dynamics and succession on different soil types.

The PPA model is a cohort model assuming time is discrete and is the following boundary value problem if the time is measured continuously (Strigul et al., 2008). The continuous version of the PPA model for m tree species consists of m von Foerster equations (1) with initial $N_i(s,0)$ and boundary conditions (2) for every species $i = 1, \ldots, m$ connected by the integral PPA equation (3) for the threshold canopy size $s^*(t)$:

$$\frac{\partial N_i(s,t)}{\partial t} = -\underbrace{\frac{\partial \left(G_i(s,s^*(t),t)N_i(s,t) \right)}{\partial s}}_{growth} - \underbrace{\mu_i(s,s^*(t),t)N_i(s,t)}_{mortality}, \tag{1}$$

$$N_i(s_{i,0},t) = \int_{s_{i,0}}^{\infty} N_i(s,t)F_i(s,s^*(t),t)ds / G_i(s_{i,0},s^*(t),t), \tag{2}$$

$$1 = \sum_{i=1}^{m} \int_{s^*(t)}^{\infty} N_i(s,t)A_i(s^*(t),s)ds, \tag{3}$$

where i indicates one of m tree species, s is the size of the tree that can be either tree height or dbh connected with height by a species specific allometric equation, $N_i(s,t)$ is the mean density of individuals of species i of size s at time t, $G_i(s,t)$ is the growth rate of these individuals i.e., $ds/dt = G(s,s^*(t),t))$, $\mu_i(s,s^*(t),t)$ is their death rate, $F_i(s,s^*(t),t)$ is their fecundity, $A_i(s^*(t),s)$ is the crown area function that gives the area of the crown at s, and $s_{i,0}$ is the size of a newborn of the ith species. Growth, death and fecundity functions depend on time t and tree size s as well as the canopy threshold level $s^*(t)$.

Strigul et al. (Strigul et al., 2008) considered transient and stationary regimes of tree monocultures as well as simple invasion and coexistence problems. The model was parameterized for different soil types so the patch (stand) dynamics at different soil and forest types types can be considered separately.

4. The Matreshka model: hierarchical scaling of forest dynamics to the landscape level.

This section introduces a modeling framework, called Matreshka (after the Russian nesting doll), for the scaling of vegetation dynamics from the individual level to the landscape level through the ecosystem hierarchical structure (Figure 7, see also Strigul et al. (2012)). The Matreshka model is a particular realization of the hierarchical patch dynamics concept (Levin & Paine, 1974; Wu & Loucks, 1996) in application to forested ecosystems. The model (Fig. 7) represents forest dynamics at the landscape level as an interference of separated processes occurring at different spatial and temporal scales: 1) within forest stands dynamics caused by individual-level disturbances, and 2) dynamics of the mosaic of forest stands caused by large disturbances. The Matreshka model can be presented as a continuous or a discrete model, where partial differential and integral equations and Markov chains are employed,

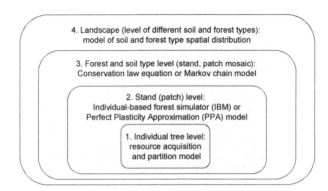

Fig. 7. The Matreshka framework for hierarchical scaling of vegetation dynamics to the landscape level

respectively. The highest hierarchical level is the landscape level comprising a mosaic of different soil and forest types. The vegetation dynamics at this level are the composition of vegetation dynamics of different forest types. The forest and soil type level consists of the mosaic of forest patches that are in different successional stages. In this model, forest patches are considered as spatial units of a considerably large size (0.5 - 1 hectare). A broad discussion of the model assumptions can be found in a recent paper (Strigul et al., 2012) focusing on the dynamics of forest stands (level 3 on Fig. 7). The Matreshka employs previously developed models for the processes at smaller scales. In particular, tree dynamics within the forest stands can be modeled by an individual-based forest growth model (for example, SORTIE, Crown Plastic SORTIE or LES models) or by forest growth macroscopic equations, specifically, the Perfect Plasticity Approximation model (PPA). The individual tree level model captures growth, mortality, and reproduction of individual trees depending on tree size, light and nutrient availability, soil type, and other factors. Several empirically determined parameters approximate these individual-level processes in the SORTIE and PPA frameworks (Pacala et al., 1996; Strigul et al., 2008).

At the next step of scaling, age-structured dynamics of forest stands (patches) on the given soil type can be described by the conservation law following Levin and Paine (Levin & Paine, 1974) in the continuous case:

$$\frac{\partial n(t,a,\xi)}{\partial t} = -\frac{\partial n(t,a,\xi)}{\partial a} - \frac{\partial g(t,a,\xi)\,n(t,a,\xi)}{\partial \xi} - \mu(t,a,\xi)\,n(t,a,\xi), \qquad (4)$$

where, $n(t,a,\xi)$, $g(t,a,\xi)$ and $\mu(t,a,\xi)$ are density, mean growth rate and extinction rate of a stand of state a and size ξ at time t, correspondingly. The initial stand distribution $n(0,a,\xi)$ and the "birth rate" of new stands should be specified to simulate given forested ecosystem.

The original model operates with two variables (patch age, a, and size, ξ), but it has been indicated (Levin & Paine (1974) p. 2745) that age is just one of the possible "physiological" variables. In this chapter, we consider a special case of equation (4), where the forest patches are fixed in size, so the rate of patch growth is zero $g(t,a,\xi) = 0$. Variable a is considered as a successional stage of forest stand, and is discussed in another paper (Strigul et al., 2012). This formulation of the Matreshka model (equations 1-4) is analytically tractable in special cases, though the general analysis is a significant challenge. In particular, in the following example we consider the stationary distribution of tree monoculture stands.

In the discrete case Strigul et al. (Strigul et al., 2012) proposed a discrete time Markov chain model for stand (patch) dynamics that can be easily generalized to a continuous time framework by taking random times between transitions. However, the discrete modeling approach has certain advantages such as that the transition of stands between stages can be explicitly defined, the probability matrix is easy to interpret and estimate using forest inventory data. In the general Markov chain model for the stand transition (Strigul et al., 2012), the states in the Markov chain are represented by stand successional stages $\{1, 2, \ldots, m\}$ characterizing the forest stand development up to a certain maturity stage m. In certain applications, such as forest fire models, the successional stage is characterized by the absolute stand age, i.e., the time since the latest major fire disturbance. However, in general, the choice of the parameter characterizing stand successional stage can be a challenging problem. The model for development of one stand (patch) may be represented using a graph as in Figure 8 and is described using a general transition probability matrix (5):

$$P = \begin{pmatrix} r_1 & p_1 & 0 & 0 & \cdots & 0 & 0 \\ q_{2,1} & r_2 & p_2 & 0 & \cdots & 0 & 0 \\ q_{3,1} & q_{3,2} & r_3 & p_3 & \cdots & 0 & 0 \\ \vdots & & & \ddots & \ddots & & \\ \vdots & & & & \ddots & \ddots & \\ q_{m-1,1} & q_{m-1,2} & q_{m-1,3} & q_{m-1,4} & \cdots & r_{m-1} & p_{m-1} \\ q_{m,1} & q_{m,2} & q_{m,3} & q_{m,4} & \cdots & q_{m,m-1} & r_m \end{pmatrix} \tag{5}$$

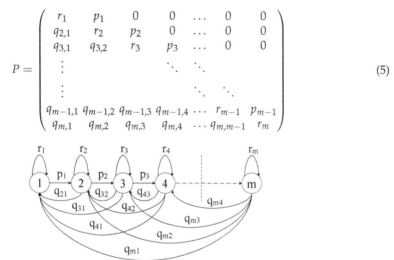

Fig. 8. A graph of the complete stage dynamics model of forest stands (after Strigul et al. (2012)).

The model assumes that the patch (forest stand) is observed frequently enough relative to the succession process so that the forest does not grow through two consecutive successional states. Each time the forest stand moves to the next stage with probability p_i or stays at the same stage with probability r_i (due to some minor forest disturbances or a small interval between forest inventories). The $\{q_{ij}\}_{i\in\{2,\ldots,m\},j\in\{1,\ldots,m-1\}}$ probabilities describe disturbances affecting stand succession. The disturbances include disaster events which completely destroy forest stands ($q_{x,1}$, $x = 2, \ldots, m$) or smaller-scale events which change the stand successional stage to one of the previous stages with certain probabilities ($q_{h,k}$, $h > k > 1$). These disturbances determine the development of forest as a mosaic of patches (stands). The model makes no distinction or explanation between the causes of the disturbances leading to the successional stage, in particular, both silvicultural operations such as forest harvesting or natural disturbances would lead to larger $q_{i,j}$ probabilities (Strigul et al., 2012).

The Matreshka model is considered as a first step in development of an analytically tractable model capable of capturing the forest dynamics on multiple scales. However, the PPA (1-3) model and the forest stand model (7) as well as its discrete counterparts such as Markov chain models (8), are only partially analytically tractable. In particular, the stationary states of these models and their stability can be relatively easily investigated, while the transient dynamics is a challenging problem. Therefore, we are still far away from complete mathematical theory of multiscale forest dynamics.

The key element for the Matreshka model is to simulate forest dynamics as a patch-mosaic phenomenon at two distinct hierarchical scales: at the individual level and the stand level (7). In forest ecology the two focal scales (i.e. individual and stand levels) have been broadly discussed with respect to forest dynamics and disturbance regimes (Bragg et al., 2004; Strigul et al., 2012). Patch-mosaic dynamics of larger forest units (stands) have also been considered in different studies, such as in forest fire models, forest disease models, and anthropogenic disturbance modeling (Bragg et al., 2004; Forman, 1995; Wu & Loucks, 1996). In the Matreshka model, we use the PPA model to scale up gap dynamics to the stand level and consider forest patches as much large spatial units (about 0.5-1 ha, see (Strigul et al., 2012) for more details).

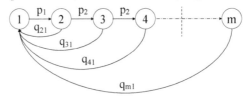

Fig. 9. A graph for a simplified forest stand model. The Birth and Disaster Markov chain.

4.1 Fire disturbance model: a case study

In this example, an analytically tractable case of the Matreshka model is considered. The simple case of the PPA model-the flat-top model-is employed to describe a tree monoculture stand (Strigul et al., 2008). The flat-top model was parameterized and validated for the Lake states (Purves et al., 2008). This model is a special case of the PPA model (equations 1-3) where tree growth and mortality are characterized by several species-specific constant parameters such as understory and overstory rates of growth as well as mortality and fecundity parameters. Using these simplest possible functional forms makes the model analytically tractable (Strigul et al., 2008). In particular, there exists a unique stable stationary state of a flat-top monoculture stand, and stationary age and size distributions of trees within the stand can be calculated. The transient dynamics are less tractable; however, the self-thinning exponents were analyzed analytically (Strigul et al., 2008), and a good approximation of the total length of transient period (t^*) for the case of the invasion into an empty habitat was derived (unpublished results). The length of the transient period curve corresponding to the stand development, starting from the invasion into an empty habitat until the stationary state, may be approximated by a piecewise linear model:

$$x(t) = \begin{cases} \alpha t, \ t \leq t^*, \\ x^*, \ t > t^* \end{cases} \qquad (6)$$

where $x(t)$ is a stand characteristic (such as biomass or cumulative basal area), x^* - stationary value of the quantity $x(t)$, t^* is the length of the transient period, and a parameter α is x^*/t^*, so it can be determined if the values x^* and t^* are known. This piecewise-linear approximation is commonly used in microbiology to approximate sigmoidal growth in microbial cultures. Sigmoidal growth models, for example, Gompertz and logistic curves, are often used to describe growth of stands and individual trees as well as microbial cultures (Dette et al., 2005; Yoshimoto, 2001).

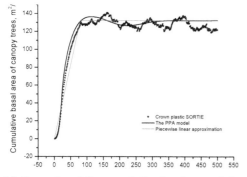

(a) Dynamics of the cumulative basal area of the hypothetical stand of white pine simulated by the crown plastic SORTIE model (black points) and the PPA model (black line) (see Strigul et al. (2008) for the details and parameter values), the red line is a piecewise linear approximation.

(b) The stationary stand age distribution of the mosaic of forest patches represented by the negative exponential distribution in the forest fire model (after Van Wagner (1978)).

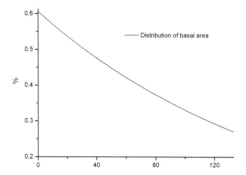

(c) The stationary distribution of stand basal area of the mosaic of forest patches.

Fig. 10. An example of the fire disturbance model for a tree monoculture.

We consider a special case of equation (4) to describe the stand level dynamics of tree monoculture. In particular, we assume that the stands are fixed in size, i.e. $g(t, a, \xi) = 0$ and have constant extinction (disaster) rate μ, to obtain the following model:

$$\frac{\partial n(t, a)}{\partial t} = -\frac{\partial n(t, a)}{\partial a} - \mu\, n(t, a). \tag{7}$$

This model describes the patch-mosaic pattern of stands, given some initial stand distribution $n(0, a)$ and assuming that new stands emerge in place of extinct stands. The discrete version of equation (7) is a birth-disaster Markov chain with constant parameters p and q (Fig. 9). This model, with the successional stage a considered as stand age, is mathematically equivalent to the classical forest fire model developed by Van Wagner (Van Wagner, 1978). It is a simple mathematical exercise to show that the model (7) has a stable stationary distribution described by a negative exponential law which after standardization can be presented as the negative exponential distribution with the following probability density function:

$$f(a,\mu) = \begin{cases} \mu e^{-\mu a} & a \geq 0, \\ 0, & a < 0. \end{cases} \tag{8}$$

The negative exponential distribution as well as its discrete version - the geometric distribution are employed in forest fire models to describe stationary age distributions of forest stands (Johnson & Gutsell, 1994; Van Wagner, 1978).

Using the Matreshka framework, we can now scale up the predictions of the PPA model to the level of mosaic of forest stands. We can invert equation (6) as a function of $t(x)$ on an interval $[0, t^*]$ and there are infinitely many values of t corresponding to the value x^*. Substituting this result in equation (8) we obtain the stationary probability distribution of the quantity x:

$$f(x,\mu,\alpha) = \begin{cases} \mu e^{-\frac{\mu x}{\alpha}} & 0 \leq x < x^*, \\ \left(1 - \int_0^{x^*} \frac{\mu}{\alpha} e^{-\frac{\mu x}{\alpha}} dx\right) \delta(x - x^*), & x = x^*, \\ 0, & x < 0 \text{ and } x > x^*, \end{cases} \tag{9}$$

where $\delta(x)$ is the Dirac delta function that accounts for all the stands which are in the stationary state. In the discrete case, the geometric distribution may be considered instead of distribution (8). In that case, the transformed distribution corresponding to (9) will have only a finite number of values. The coefficient for the resulting Dirac delta function in (9) will be the last value corresponding to x^* in this distribution.

As an illustrative example, we consider a stand of white pine (*Pinus strobus*) simulated by the crown plastic SORTIE and the corresponding PPA model. Figure 10(a) presents the simulation results (reproduced with permission from (Strigul et al., 2008)). The model functional forms and parameter values are available in the latter reference. In this example, the parameter $x(t)$ is a stand cumulative basal area; however, biomass, average canopy height etc. may be employed instead. Figure 10(b) illustrates the negative exponential distribution of stand ages corresponding to the stationary state of equation (7) with $\mu = 0.01$. This parameter value corresponds to an example considered by Van Wagner in his classical work on forest fire modeling (Van Wagner, 1978). Figure 10(c) presents the distribution (9), where 44.93% of all stands have the stationary state basal area $x^* = 132 \, m^2/ha$. Note that the shape of the distribution (9) is determined by the values of μ, x^*, and t^*. Therefore, the stationary distribution (9) predicted by this simple modification of the Matreshka model may be observed and verified subject to data availability.

This example is based on the tree monoculture model (Strigul et al., 2008) and therefore is of limited practical value for the SFM of indigenous multispecies forests. However, even this simplified model can be implemented directly for certain forest types that are naturally dominated by one tree species. One particular example is the longleaf pine (*Pinus palustris* Mill.) forest that has historically dominated the Southeastern United States. This natural monoculture ecosystem was supported by forest fires, as the longleaf pine is fire resistant. Development of its competitors, such as loblolly (*Pinus taeda*) and slash (*Pinus elliottii*) pines, has been limited by frequent forest fires. Over the last 150 years the landscape has changed radically due to overexploitation and fire suppression. Intensive longleaf pine forest restoration projects at the Southeastern U.S. are currently on-going within the SFM framework (www.longleafalliance.org).

This example demonstrates the potential advantages of using the multiple-scale modeling for SFM applications. The Van Wagner fire-disturbance model operates with the stand age after a major disturbance (Van Wagner, 1978). Therefore, the forest management plans within this model should be based on the fire-disturbance history. In practice, the exact fire history is often hard to determine. Forest surveys, such as USDA FIA data and Canadian forest service data, determine the stand age empirically as an average age of canopy trees. This parameter is unfortunately not very reliable for modeling purposes (Strigul et al., 2012). The Matreshka model allows one to develop the forest management plans using forest stand stratification with respect to the stand successional stage, basal area, or stand biomass. The stand biomass or basal area can be easily calculated using available survey data (Strigul et al., 2012) and the forest management plan can be designed based on these stand characteristics. This makes the model suitable for the needs of criterion 1.1 (Ecosystem diversity) of the Montreal process.

4.2 Application of the Matreshka model to criterion 5 of the Montreal process
The Matreshka model is developed for ecological forestry and SFM applications. Specifically, it allows one to incorporate natural and anthropogenic disturbances occurring at different scales, ranging from individual trees to stands to predict forest growth at the landscape level. To address criteria 2 and 5 of the Montreal process, the model can naturally incorporate effects of climate change on individual tree growth through modification of either the forest individual-based model or the PPA model. Changes of the natural disturbance regime due to climatic factors can be incorporated by modification of tree mortality functions or by changing elements and structure of the transition matrix 5 (for stand-level disturbances). Similarly, changes in forest policy, silvicultural practices, and anthropogenic disturbances can also be incorporated in the model through modification of tree mortality functions and the transition matrix 5. While the Matreshka model is formulated as a non-spatial model at the stand level, the model can also be presented in a spatially explicit form by using GIS-based simulations of forest stands at the landscape level. This can be essential if the forest stewardship in the focal area varies due to different landowner policies.

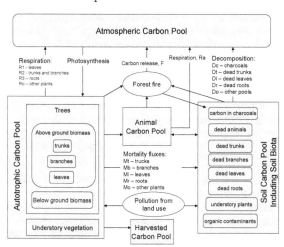

Fig. 11. The framework for modeling of forest carbon footprint for the SFM applications. The Matreshka model is used for the modeling of Autotrophic and Soil Carbon Pools.

Current on-going research is focused on the application of the Matreshka model to the carbon cycle modeling of temperate forests in the North-Eastern Part of the USA and Quebec in agrement with criterion 5 of the Montreal process "Maintenance of forest contribution to global carbon cycles". The carbon footprint of forest ecosystems is determined by the dynamics of carbon sequestration and release, and can be affected by harvesting and other anthropogenic activities. In this project, the Matreshka model is used to predict the forest carbon cycle according to a conceptual model presented in Figure 11. Most of the carbon influx into the ecosystem is derived from photosynthetic assimilation of atmospheric CO_2 by the autotrophs (overstory trees, understory trees, shrubs, and groundcover vegetation) that determine the gross primary productivity (GPP). The major effluxes of carbon in the atmosphere occur as the result of autotrophic respiration (which is defined as the sum of maintenance respiration and growth respiration), heterotrophic respiration, and the processes of physical decomposition of organic matter, such as fire. Carbon is also removed from a forest ecosystem by wood harvesting. The typical parameters of interest in calculating carbon footprints are the net primary production (NPP, defined as GPP minus autotrophic respiration), and the net ecosystem production (NEP). The NEP is determined as the net exchange of CO_2 between the atmosphere and ecosystem, which is measured on an annual basis, and equal to the NPP minus heterotrophic respiration. During recent decades, carbon fluxes presented in Figure 11 were evaluated, however, the current models operate with the carbon balance at the macroscopic level using average estimates of the carbon pools and fluxes. In this project, the key element for predicting forest carbon cycle is the Matreshka model. This model provides a scaling of carbon balance from an individual tree level to the stand level, and simulates the autotrophic carbon pool (Fig. 11). Therefore, the carbon balance model incorporating the Matreshka model scales up the effects of the silvicultural practices and other anthropogenic activities from the individual tree-based level to the ecosystem level, and can predict changes in the forest structure and carbon dynamics at different time horizons.

5. Conclusion

In this chapter the framework of complex adaptive systems is employed to address the basic challenge of the ecological forestry and SFM, i.e., to understand and predict how natural and anthropogenic disturbances occurring at different scales propagate through the forested ecosystems and affect forest structure and dynamics. This framework naturally combines experimental and theoretical approaches. This framework consists of three major components: 1) the development of individual-based models (IBMs) to simulate multiple scales processes in complex systems, and their parameterization with experimental data (in particular, by using USDA forest inventory data, FIA); 2) the development of different scaling methods that approximate individual-based processes; and 3) validation with real data and practical applications. The first component involves mostly computer simulations of what are, in general, analytically-intractable stochastic processes. Forest growth IBM can serve as an intermediate research step in the derivation of macroscopic equations (i.e., tractable analytic models approximating this stochastic process), and, as an independent research tool, to simulate forest carbon balance, stand dynamics, natural disturbances (such as disease outbreaks), and the outcomes of silvicultural prescriptions. Scaling methods may allow models to be reduced to analytically tractable objects, macroscopic equations-such as stochastic and deterministic dynamical systems-which are both more robust in their predictions and, also, computationally simpler. Recently developed models including the Crown Plastic SORTIE, LES, and PPA have been developed within this research framework to address the scaling of vegetation dynamics from the individual to the stand level. All these models employ individual tree plasticity as a crucial factor for canopy development and forest self-organization within the stand level. The Matreshka model generalizes these models operating on the individual level for scaling of vegetation dynamics to the landscape level using the hierarchical patch dynamics concept. It is anticipated that these new modeling tools will be employed for the SFM of indigenous forests. Practical applications of the developed modeling approach address criteria 2 and 5 of the Montreal process. The ongoing research focuses on the modeling of the temperate forest carbon cycle in the North-Eastern USA and Quebec. The Matreshka modeling framework can help natural resource managers to understand how changes in forest management practices can affect the forest carbon footprint, and to manage the key ecosystem processes that control carbon and nutrient dynamics in a forest ecosystem.

6. Acknowledgement

I would like to thank Simon Levin and Stephen Pacala for inspiring my research. I would like also to acknowledge my students Alicia Welden, Ian Cordasco and Fabian Michalczewski who participated in this project at various times.

7. References

Aaltonen, V.T. (1926) On the space arrangement of trees and root competition. *J For* 24:627-644.

Assmann, E. (1970) *The principles of forest yield study.* Pergamon Press, Oxford.

Ballaré, C.L. (1999) Keeping up with the neighbours: phytochrome sensing and other signalling mechanisms. *Trends Plant Sci* 4:97-102.

Biging, G.S. & Dobbertin, M. (1995) Evaluation of competition indices in individual tree growth models. *For Sci* 41:360-377.

Botkin, D.B. (1993) *Forest dynamics, an ecological model.* Oxford University Press, New York.

Bragg, D.C., Roberts, D.W. & Crow, T.R. (2004) A hierarchical approach for simulating northern forest dynamics. *Ecol Model* 173:31-94.

Brisson, J. (2001) Neighborhood competition and crown asymmetry in *Acer saccharum. Can J For Res* 31:2151-2159.

Bugmann, H. (2001) A review of forest gap models. *Climatic Change* 51: 259-305.

Burton, P.J. (1993) Some limitations inherent to static indices of plant competition. *Can J For Res* 23: 2141-2152

Busgen, M. & E. Munch. (1929) *The structure and life of forest trees.* Chapman and Hall, London

Casper, B.B, Schenk, H.J. & Jackson R.B. (2003) Defining a plant's below-ground zone of influence. *Ecology* 84:2313-2321.

Clark, J.S. (1991) Disturbance and tree life history on the shifting mosaic landscape. *Ecology* 72:1102-1118.

Cole, W.G.& Lorimer, C.G. (2005) Probabilities of small-gap capture by sugar maple saplings based on height and crown growth data from felled trees. *Can J For Res* 35:643-655.

Curtin, R.A. (1970) Dynamics of tree and crown structure in *Eucalyptus oblique. For Sci* 16:321-328.

Dette, H., Melas, V.B. & Strigul, N.S. (2005) Application of Optimal Experimental Design in Microbiology. in *Applied Optimal Designs*, M. Berger and W.K. Wong (Eds), Willey, 137-180.

Engler, A. (1924) Heliotropismus and geotropismus der beume und deren waldbaumliche bedeutung. *Mitteilungen der Schweizerischen Centralanstalt fi£¡r das Forstliche Versuchswesen* 13: 225-283

Firn, R.D. (1988) Phototropism. *Biol J Linn Soc* 34:219-228.

Ford, E.D. (1992) The control of tree structure and productivity through the interaction of morphological development and physiological processes. *Int J Plant Sci* 153: S147-S162.

Forman, R.T.T. (1995) *Land Mosaics: The Ecology of Landscapes and Regions.* Cambridge University Press, Cambridge, NY

Franklin, J.F., Mitchell, R.J. & Palik, B.J. (2007) *Natural disturbance and stand development principles for ecological forestry.* Gen. Tech. Rep. NRS-19. Newtown Square, PA: U.S. Department of Agriculture, Forest Service, Northern Research Station.

Frelich, L.E. & Martin, G.L. (1988) Effects of crown expansion into gaps on evaluation of disturbance intensity in northern hardwood forests. *For Sci* 34: 530-536.

Godin, C. (2000) Representing and encoding plant architecture: A review. *Ann For Sci* 57: 413-438.

Grace, J. (1977) *Plant response to wind.* Academic Press New York, USA.

Grosenbaugh, L.R. (1981) Measuring trees that lean, fork, crook, or sweep. *J For* 89-92.

Grote, R. & Pretzsch, H. (2002) A model for individual tree development based on physiological processes. *Plant Biol* 4: 167-180.

Gysel, L.W. (1951) Borders and openings of beech-maple woodlands in southern Michigan. *J For* 49:13-19

Harker, R.I. (1996) Curved tree trunks - Indicators of soil creep and other phenomena. *J Geol* 104:351-358.

Hibbs, D.E. (1982) Gap dynamics in a hemlock-hardwood forest. *Can J For Res* 12:522-527.

Holbrook, N.M. & Putz, F.E. (1989) Influence of neighbors on tree form: effects of lateral shade and prevention of sway on the allometry of *Liquidambar styraciflua* (Sweet Gum). *Am J Bot* 76:1740-1749.

Ishii, R. & Higashi, M. (1997) Tree coexistence on a slope: An adaptive significance of trunk inclination. *Phil Trans R Soc Lond B* 264:133-139.

Ishizuka, M. (1984) Spatial pattern of trees and their crowns in natural mixed forests. *Jap J Ecol* 34:421-430.

Johnson, E.A. & Gutsell, S.L. (1994) Fire frequency models, methods and interpretations. *Adv Ecol Res* 25:239-283.

King, D.A. & Loucks, O.L. (1978) The theory of tree bole and branch form. *Radiat Environ Bioph* 15: 141-165.

Kleunen, M. & Fischer, M. (2005) Constraints on the evolution of adaptive phenotypic plasticity in plants. *New Phytol* 166: 49-66.

Krajicek, J.E., Brinkman, K.A. & Gingrich, S.F. (1961) Crown competition-a measure of density. *Forest Sci* 7:35-42.

Lane-Poole, C.E. (1936) Crown ratio. *Aust For* 1:5-11.

Lawrence, D.B. (1939) Some features of the vegetation of the Columbia River Gorge with special reference to asymmetry in forest trees. *Ecol Monogr* 9:217-257.

Levin, S.A. (1999) *Fragile dominion: complexity and the commons.* Perseus Publishing, Cambridge, MA.

Levin, S.A. (2003) Complex adaptive systems: Exploring the known, the unknown and the unknowable. *B Am Math Soc* 40:3-19.

Levin, S.A. & Paine, R.T. (1974) Disturbance, patch formation, and community structure. *Proc Nat Acad Sci USA* 71(7):2744-2747.

Liu, J. & Ashton, P.S. (1995) Individual-based simulation models for forest succession and management. *Forest Ecol Manag* 73:157-175.

Loehle, C. (1986) Phototropism of whole trees: Effects of habitat and growth form. *Am Midl Nat* 116: 190-196.

Loehle, C. (1997) The adaptive significance of trunk inclination on slopes: a commentary. *Phil Trans R Soc Lond B* 264:1371-1374.

Macdonald, E. & Hubert, J. (2002) A review of the effects of silviculture on timber quality of Sitka spruce. *Forestry* 75:107-138.

McCarthy, J. (2001) Gap dynamics of forest trees: A review with particular attention to boreal forests. *Environ Rev* 9:1-59.

McMahon, T.A. & Kronauer, R.E. (1976) Tree structures: deducing the principle of mechanical design. *J Theor Biol* 59:443-66.

Mitchell, K.J. (1969) Simulation of growth of even-aged stands of white spruce. *Yale Univ Sch For Bull* 75:1-48.

Mitchell, K.J. (1975) Dynamics and simulated yield of Douglas Fir. *For Sci Monogr* 17:1-39.

Mitchell, K.J. (1980) Distance dependent individual tree stand models: concepts and applications. p. 100-137 in K.M. Brown and F.R. Clarke, edts. *Forecasting forest stand dynamics: proceedings of the workshop,* June 24, 25, Lakehead Univ., Canada.

Nyland, R.D. (1996) *Silviculture concepts and applications.* New York: McGraw-Hill Co., Inc.

Olesen, T. (2001) Architecture of a cool-temperate rain forest canopy. *Ecology* 82:2719-2730

Oliver, C. & Larson, B. (1996) *Forest stand dynamics.* New York: John Wiley & Sons, Inc.

Pacala, S.W., Canham C.D., Saponara, J., Silander, J.A., Kobe, R.K. & Ribbens, E. (1996) Forest models defined by field measurements: estimation, error analysis and dynamics. *Ecol Monogr* 66:1-43.

Perry, D.A. (1998) The scientific basis of forestry. *Ann Rev Ecol Syst* 29:435-466.

Purves, D., Lichstein, J., Strigul, N.S. & Pacala, S.W. (2008) Predicting and understanding forest dynamics using a simple tractable model. *P Natl Acad Sci USA* 105(44):17018-17022

Reventlow, C.D.F. (1960) *A treatise on forestry.* Society of Forest History. Sweden

Rouvinen, S. & Kuuluvainen, T. (1997) Structure and asymmetry of tree crowns in relation to local competition in a natural mature Scots pine forest. *Can J For Res* 27:890-902.

Runkle, J.R. & Yetter, T.C. (1987) Treefalls revisited: Gap dynamics in the Southern Appalachians. *Ecology* 68:417-424.

Ryel, R. J. & Beyschlag, W. (2000) Gap dynamics. Pages: 251-279 in: B. Marshall, and J. A. Roberts, editors. *Leaf Development and Canopy Growth.* Sheffield Academic Press.

Shugart, H.H. (1984) *A theory of forest dynamics. The ecological implications of forest succession models.* Springer, New York, USA.

Smith, D.M., Larson, B.C., Kelty, M.J. & Ashton, P.M.S. (1997) *The practice of silviculture: applied forest ecology.* Wiley, New York.

Stoll, P. & Schmid, B. (1998) Plant foraging and dynamic competition between branches of *Pinus sylvestris* in contrasting light environments. *J Ecol* 86:934-945.

Strigul, N.S., Pristinski, D., Purves, D., Dushoff, J. & Pacala, S.W. (2008). Scaling from trees to forests: Tractable macroscopic equations for forest dynamics. *Ecol Monogr* 78:523-545.

Strigul, N.S., Florescu, I., Welden, A.R. & Michalczewski, F. (2012). Modeling of forest stand dynamics using Markov chains. *Environ Model Soft* 31: 64-75.

Takenaka, A. (1994) A simulation-model of tree architecture development based on growth-response to local light environment. *J Plant Res* 107:321-330.

Trimble, R. & Tryon. H. (1966) Crown encroachment into openings cut in Appalachian hardwood stands. *J For* 64:104-108.

Umeki, K. (1995a) Modeling the relationship between the asymmetry in crown display and local environment. *Ecol Model* 82:11-20.

Umeki, K. (1995b) A comparison of crown asymmetry between Picea abies and Betula maximowicziana. *Can J For Res* 25:1876-1880.

Van Wagner, C.E. (1978) Age-class distribution and the forest fire cycle. *Can J Forest Res* 8(2):220-227.

Waller, D.M. (1986) The dynamics of growth and form. Pages 291-320 in M. J. Crawley, editor. *Plant Ecol* Blackwell Scientific Publications, Oxford.

Webster, C.R. & Lorimer, C.G. (2005) Minimum opening sizes for canopy recruitment of midtolerant tree species: A retrospective approach. *Ecol Appl* 15:1245-1262.

Westing, H. & Schulz, H. (1965) Erection of a leaning eastern hemlock tree. *For Sci* 11:364-367.

Woods, F.W. & Shanks, R.E. (1959) Natural replacement of chestnut by other species in the Great Smoky Mountains National Park. *Ecology* 40:349-361.

Wu, J. & Loucks, O.L. (1996) From balance of nature to hierarchical patch dynamics: A paradigm shift in ecology. *Quart Rev Biol* 70 (4):439-466.

Yoshimoto, A. (2001) Application of the Logistic, Gompertz, and Richards growth functions to Gentan probability analysis. *J For Res* 6:265-272.

Young, T.P. & Hubbell, S.P. (1991) Crown asymmetry, treefalls, and repeat disturbance in a broad-leaved forest. *Ecology* 72:1464-1471.

8

Decision Support Systems for Forestry in Galicia (Spain): SaDDriade

Manuel Francisco Marey-Pérez, Luis Franco-Vázquez,
and Carlos José Álvarez-López
Universidad de Santiago de Compostela
Spain

1. Introduction

Ever since they were created in the 70s, Decision Support Systems (DSS) have been a great source of help with different management problems such as the optimization of travel times in airlines or train companies, medical diagnosis, business management, natural resource management, agriculture and forestry. Forest planning uses forest simulators that usually include growth and performance models in order to generate the different alternative management programs which will give rise to different production processes and programs. The selection of the best choice according to predefined criteria, which are generally related to the number of alternative management options, production programs and assessment criteria, require the use of optimization methods. These optimization methods range from whole linear programming, goal programming to heuristic methods, among which tabu search, genetic algorithm and simulated annealing are worth noting.

1.1 Development and current situation of forest DSS

In recent years, the aims for the development of forest DSS have changed. They used to have only one objective, which was to provide information about: Site index for reforestation (Hackett and Vanclay, 1998); soil fertility (Louw and Scholes, 2002); habitat requirements (Store and Jokimäki, 2003); tree growth (Hackett and Vanclay, 1998); forest management (Kolström and Lumatjärvi, 1999); wildfires (Kaloudis 2005 y Bonazoutas et al. 2008); trees brought down (Mickouski et al. 2005; Olofsson and Blennow, 2005; Zeng et al. ,2007); seed bank long-term planning (Nute et al. 2005 y Twery et al. 2005); river flow and its relation with trees (MacVicar et al., 2007) and profits (Huang et al. 2010).

More recent DSS methods have several aims. the management planning problem is focused on two or more objectives, some of which may be in conflict (Stirn, 2006). Næsset (1997) states the need for new tools to help planning aims related to biodiversity and wood production. The needs of sustainable forest management must be considered and included in the design of DSS for forestry (Wolfslehner and Vacik, 2008). Moreover, forest management actions cannot be considered in isolation or with just one objective in mind, despite the difficulties of integrating them spatially and temporally (Kangas and Kangas y MacMillan and Marshall, 2004). The aim is to offer a general view of alternative focus to face the uncertainty from the perspective of forestry and natural resource and ecosystem management (Rauscher (2000).

Some remarkable systems developed with multiple aims in forestry and natural resources are those developed for: hydrographic basins in Australia by Bryan and Crossman (2008); carbon in Canada by Kurz et al. (2009); plague control in Poland by Strange et al. (1999); landscape management in South Carolina by Li et al. (2000); visualization tool for landscape valuation by Falcão et al. (2006).

The prediction of forest activity future behaviour in its multiple dimensions is inherent to the main aim of forest DSS, therefore, the temporal scale is specially important in the development of this type of applications. Temporal scales are classified into three types: strategic, tactic and operational.

Long-term management planning or strategic is that which has a planning horizon of more than 15 years. Potter et al. (2000) state that forest ecosystem management implies the need to forecast the future state of complex systems, which often experience structural changes. It is by means of strategic planning that ecological integrity and sustainability (Gustafson y Rasmussen, 2002), risk management (Borchers, 2005 y Heinimann, 2010) and future landscape (Aitkenhead and Aalders, 2009) are guaranteed. Næsset (1997) stresses the importance of the integration of GIS with quantitative models for long term forest management. Some interesting examples of strategic forest DSS are those presented by Boyland et al. (2006) for a planning horizon of 250 years; Wolfslehner and Vacik (2008) for 120 years; Díaz-Balteiro and Romero (2004) for 100 years divided into periods of ten; Baskent et al. (2001) for 85 years; Huth et al. (2005) for 60 years and Lasch et al. (2005) for 50-year simulations.

In the case of strategic planning, it is not only the temporal scale that is higher, but also the spatial scale (Kangas and Kangas, 2005).

When the planning period is between 1 and 15 years long, it is called tactical planning or mid-term. Kangas and Kangas (2005) point that in tactical planning the number of alternative forest plans can be considered infinite. Different examples of this type of planning are those presented by Anderson et al. (2005) and Snow and Lovatt (2009). Anderson et al. (2005) present FTM (Forest Time Machine), which simulates the development of a forest area and calculates the stand development in five-year intervals. Snow and Lovatt (2008) examine the use of the general planner for agro-ecosystem models (GPAM) in pasture rotation length, building a decision tree.

Short-term or operational planning is that whose planning period lasts for a maximum of a year. Acuña et al. (1997) remark the usefulness of transparent, operative, easily validated processes provided by experts. Ducey and Larson (1999) state that sustainability assessment requires a careful balance between short-term and long-term goals. Mowrer (2000) considers that the temporal scale on which the operative tool or DSS works will have an effect on the level of uncertainty of the analysis. Uncertainly is lower in short-term planning. The more empirical the models, the more accurate they will be (Porté and Bartelink, 2002). Some remarkable examples of operative forest DSS are: Vacik and Lexer's (2001) which assesses nine species mixes and seven multiobjective regeneration methods at stand level. Newton (2003) tabulates the annual management of spruce plantation. Thomson and Willoughby (2004) present a web system of forest management consultancy. Kurz et al. (2009) comment a version on operative scale of the model of carbon-dynamics. Newton (2009) shows the usefulness of the modular-based structural stand density management model (SSDMM) for decision-making in operational management.

Another relevant aspect in forest DSS analysis is the spatial scale of the unit of analysis. The highest level is on a regional or, in some cases, national scale. In this type of works planning is strategic and establishes the guidelines (Anderson et al. 2005; Ascough, 2008; Carlsson et al. 1998; Crookston and Dixon, 2005; Heinimann, 2010; Kurz, 2009; Mathews,1999; Mowrer, 2000, Potter, 2000; Maitner et al. 2005; Nute, 2005; Reynolds, 2005; Thompson et al. 2007). These works have been developed mainly in the USA and in different regions in Scandinavia.

A second scale is the lanscape or forest that has been the unit of analysis for the studies carried out in the USA (Rauscher in 1999, Twery and Thomson et al. in 2000; Twery and Hornbeck in 2001, Bettinguer et al. 2005; Borchers, 2005; Gärtner et al. 2008 and Graymore et al. 2009). In Scandinavia some remarkable studies are those by Anderson et al. 2005; Kurttila, 2001; Leskinen et al. 2003; Store and Jokimäki, 2003, among others. In other countries, the works by Seely et al. 2004; Stirn, 2006 and Wang et al., 2010 are worth noting. It is at this scale when the need to integrate GIS within DSS arises. It should be easy for regional managers to carry out forest zoning effectively in the areas where initiatives are necessary for sustainable progress (Martins and Borges, 2007). It is also necessary to integrate 3D visualization tools (Falcão et al., 2006).

The main planning scale so far is the stand, where units are homogeneous regarding ecology, physiography and future developments. Some remarkable works are those by Aerstenet et al. 2010; Anderson et al. 2005; Crookston and Dixon, 2005; Ducheyne et al. 2004; Huth et al. 2005; Kolström and Lumatjärvi, 1999; Mathews et al. 1999; Mette et al. 2009; Torres-Rojo and Sanchez-Orois, 2005; Seely et al. 2004; Snow and Lovatt, 2008; Twery et al., 2000 and Varma et al. 2000. They have been applied in different areas such as Scandinavia, Australia, Austria, Canada, Malaysia, Scotland, Germany and Turkey for mixed stands of different species such as firs, spruces and tropical species among others. Works on this scale with differenciating characteristics are those by Baskent et al. (2001) about simulated stands; by Vacik and Lexer (2001) and Kurttila (2001) applied to stands from natural regeneration; and those by Chertov et al. (2002) and Goldstein et al. (2003) which analyze the consequences on natural ecosystems. The studies by Nute et al., (2005), Twery et al. (2005) and Salminen et al. (2005) enable the user to update the investment assessment on a stand level and on a whole exploitation level, developing thus scenarios of one or more treatments for management units. Martins and Borges (2007) point out that the search for sustainability of woods belonging to a high number of non-industrial private forest owners (NIPF owners) requires devising tools of the appropriate size for the properties and decision scale.

Flexibility in decision-making has become an essential element in the development of forest DSS in recent years. There has been a development from methods that allowed only unilateral decisions, only one person has the decision-making power (Thomson et al. 2000; Leskinen et al. 2003; Kaloudis et al. 2005). Even if unilateral systems have been maintained , new ones have been developed where decision is collegial, that is, multiple participants express their preferences to support an only actor in the decision-making. In Kangas and Lekinen (2005), some experts choose the explicative variables that will be used in the model after a careful study of the forest area. The software for damage reduction by fire proposed by Kaloudis et al. (2010) has been initially tested and evaluated by three different groups of users.

The most recent, interesting and complex issue in forest DSS are those with participative decision-making by several stakeholders who must reach an agreement for a final decision. In Nute et al. (2000) decision-making is developed with a social participative and environmentally sensitive methodology in Central America. Mendoza and Prabhu (2003) use MCA methodology to carry out an assessment of the Criteria and Indicators (CandI) structure in an environment of participative decision-making. In a context of public participation, Sheppard and Meitner (2005) describe the managers needs in sustainable forest planning, outlining the criteria for the design of support processes for these decisions. According to Mendoza and Martins (2006) the qualitative method allows a more participative decision-making process. In public participation processes, Kangas et al. (2006) state the importance of questions such as equity, representativity and transparency. Martins and Borges (2007) interprets the design of a forest management plan as a case of participative planning. Ramakrishnan (2007) uses participative management methods in sustainable forestry. Other illustrative examples are those presented by Vainikainen et al. and Wolfslehner and Vacik in 2008 and Anderson et al. in 2009.

It is important to highlight the development of mathematical tools and their implementation by means of information technology for efficient problem solving. Díaz-Balteiro and Romero's work entitled "Making forestry decisions with multiple criteria: A review and an assessment" (2008), makes an excellent assessment of the different issues in forest management and the different problem-solving tools. Finally, there are some references that haven't been used in the afore mentioned work and that have incorporated some different types of relevant techniques for forest DSS: Aitkenhead and Aalders (2009) use Bayesian networks; Martín-Fernández and García-Abril (2005) and Zeng et al. (2007) use genetic algorithms and tabu searches; Stirn (2006) uses dynamic programming and fuzzy techniques; Chertov et al. (2002) use data mining; Wolfslehner et al. (2005) use AHP and ANP; and MacMillan and Marshall (2004) use lineal programming.

2. Forestry sector in Galicia¶

The forestry sector is crucial in Galicia from a strategic, social and economic point of view, Figure 1 (Marey-Pérez and Rodríguez-Vicente, 2008). In recent years, Galicia has produced half the wood in Spain, becoming in some periods the ninth country in the wood harvest rank in the EU, even above the United Kingdom (FEARMAGA, 2009). In the past ten years, forest producers in Galicia have perceived 1,000 million euros due to wood selling. Due to this production capacity, Galicia has a wood transformation sector with around 3,500 companies, mostly family business, which employs 26,000 people directly and 50,000 indirectly. In fact, it is overall the third most important industrial activity and in twenty out of the fifty-six forest regions it is either the first or the second. These forest regions are located mainly in rural environments, so it is one of the main assets for the sustainability of rural population (Marey-Pérez and Díaz-Varela, 2010).

One of the most serious problems for this sector is its atomization (Marey-Pérez et al. 2006). This stems from the small size of the property of each owner and of each parcel. This results into 700,000 forest owners with less than 3 ha of average property divided into more than 8 parcels (Rodríguez-Vicente and Marey-Pérez, 2010).

3. A supporting decission forest system: SaDDriade

3.1 Origins

New information and communication technologies are currently, and will be to a higher extent in the future, the basic pillar on which the economic development of our society will lie. In rural areas in Europe the currently-existing digital divide will be overcome by different means: public funding, training and the initiatives of organizations and companies

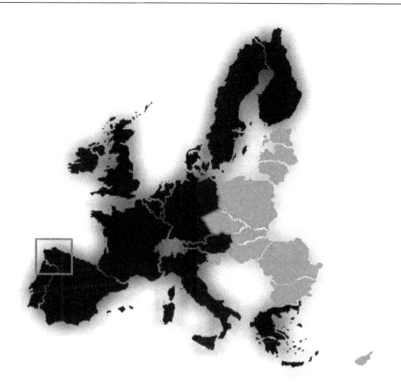

Fig. 1. Location of Galicia in Europe and Spain

regarding these types of activities. These initiatives are going to depend on the DSS problem-solving capacity.

In the Research group "Proyectos y Planificación de la Universidad de Santiago de Compostela (GI-1716)" we have been working on solutions to the problems experienced by agricultural and agroindustrial sectors in Galicia and in other rural areas in Europe and Latin America. SaDDriade is the result of a process that started with the analysis of the weaknesses and strengths of the forestry sector. Strategic plans regarding forest industry were revised, wood, furniture, energy, environmental preservation and land planning, among others. Special emphasis was placed on the revision of information technologies and sustainable development with the idea of gathering as much information as possible. The experience of the research group was incorporated to the corpus of knowledge, providing it with a scientific and practical dimension.

Symposia, conferences, forums and meetings with owners associations, industry, administrations and scholars have been organized in recent years. This provided very valuable reflections for the definition of a DSS adapted to the reality of our region. Finally, financial, technical and scientific support to start the project has been obtained as a result of the combination of the Xunta de Galicia research Project "*Sistema de apoio a decisión para montes veciñais en man común (SadMvmc)*" *(07MRU035291PR)* and the different collaborations with public administrations, forest associations and private companies within the framework of the project *COST Action FP0804 - Forest Management Decision Support Systems (FORSYS).*

3.2 Motivations

The study of the data gathered provided the keys about the demands that SADDriade should answer and also of the way in which this should be done to provide the right answers to potential users. Below, there is a selection of weak points of different aspects within the forestry sector from different reports chosen due to their different degrees of usefulness for the development of the system.

- Make information available to facilitate strategic decision-making by the companies in the forest chain and guide the definition of public policies.
- Promote the activities of support services (researcher training or services to companies) that encourage industries' main activities.
- Decrease the lack of appropriateness of the formative level of the general population to the requirements of the companies. Despite the recent efforts and advances, there is still an imbalance between the needs of the companies and the development of technologies.
- Collaborate in a rational exploitation and use of natural spaces and in the reduction of industrial impact on the ecosystem.
- Encourage the collaboration of universities, technological centers and companies.
- Promote forest as a source of income for companies and private owners. Forest smallholder property poses difficulties for sustainable developments since forest areas are considered a secondary source of income and specific forestry is not developed.
- Provide businessmen with competitive advantages in tangible resources such as financial, technological or natural resources to counteract the competence in Price of markets with cheaper labour.
- Promote the implementation of technology in companies where the presence of technology is not enough, which limits their competitivity and the development of forest activities.

In our proposals, our aim was to make a forest DSS of immediate use for companies. It would have no cost for them in implementation and in licenses, and it would not require specific training for its users or an equipment update. It would also reach the highest possible number of users in the shortest possible time, which made it necessary to have a fluent transmission of knowledge. Our experience as university instructors enables us to identify the most efficient means in which users acquire knowledge. After considering different possibilities, we reached the conclusion that the only option that fulfilled all the requirements was the world wide web, using a web application.

Within the creation process of a software tool, the choice of a certain development platform is a key issue that conditions the rest of the actions. It is necessary to make a detailed analysis of the weaknesses and strengths of each programming environment and compare them with potential user profiles and the requirements for a satisfactory user experience. Currently, the technological developments and the increase in telecommunications favours the development of web applications and their merge with mobile ones, instead of with desktop applications, which are becoming less important. Some of the reasons to choose a web platform are:

- Multi-platform compatibility: Several technologies such as PHP, Java, Flash, ASP and Ajax allow an effective development of programs supporting the main operating systems.
- Update: Being always updated without pro-active user actions and without calling the user's attention or interfere in his/her working habits.

- Immediacy of access: Web-based applications don't need to be downloaded, installed or configured. They can be accessed via an internet address and be ready to work, regardless of their configuration or hardware.
- Easy trial: There are no obstacles for easy and effective tool and application trials.
- Lower memory requirement: They have more reasonable RAM memory requirements for the end-user than programs installed locally. Since they are stored and run in the provider's servers these web applications use these servers' memory in many cases.
- Fewer errors: They are less likely to 'freeze' and cause technical problems due to hardware conflicts with other existing applications, protocols or internal personal software. In web applications, they all use the same version and all the errors can be corrected as soon as they are found.
- Price: They don't require the distribution infrastructure, technical support and marketing required by traditional desktop software.
- Multiple concurrent users: These applications can be used by multiple users at the same time. There is no need to share screens when multiple users can jointly see and even edit the same document.
- Usability: Web browsers are by far the computer application with the highest presence in computers around the world due to the expansion of internet as a channel of communication for businesses and particulars.

3.3 Who made SaDDriade?

The SaDDriade team is made up of ten people with the collaboration of different professionals and technicians from the forestry sector, the administration and more than twenty well-known international experts in the field.

Gradually, technical-economical models have been built for the different forest species and their different locations in Galicia. These models include the most advanced techniques in forestry and individual tree or forest growth, together with the parametrized financial component of the different forest management phases or tasks: land preparation for planting, tasks linked to forestry and the use of wood or biomass.

3.4 Users of SaDDriade

During the development of all its components, we considered the potential users, their demands and their previous knowledge. In this way, we have developed a user-friendly, accessible, usable application, with a clear presentation of results. It also enables the user to determine in which phase of the work he/she is without being trapped within the program. The characteristics of the potential clients and the demands that SaDDriade answers are outlined below:

- Forestry business: Companies related to reforestation, forestry, wood and forest biomass harvest, first transformation industry (sawmills, wood plank industry, paper and pulp industry and biomass power generation) and consultancies and engineering technical offices that advise owners.
- Forest owner associations: their needs and demands are very similar to those of the previous group. However, most of them lack the qualified technical staff to rigorously go through the planning process. SaDDriade has been conceived with the idea of helping them improve the quality and quantity of their forest product.
- Research groups in universities and research institutes: SaDDriade can perform the role of a lab and database with useful information to understand how forest reality works.

All this knowledge and experience will result into new studies that will contribute to quality forestry based on multifunctionality, energetic use and rational resource planning. Its ultimate aim is the excellence and sustainability of forest farms that will secure the financial future of small forest farms in short, middle and long term.

- Forest training centers. Forest training centers and specially those departments at universities devoted to forest activity can use this application for teaching. Students can acquire experience in forest planning and management and use forest simulators in their area seeing the possibilities of evolution and acquiring practical knowledge that would be impossible otherwise.

3.5 Objectives of SaDDriade

The main objective is to provide information about the productive cycles of the different forest species in Galicia. The data provided are going to enable the knowledge and assessment of the different steps to be taken in the productive processes, the costs associated to each of them and the expected final yield.

Users receive answers to queries regarding different aspects of forest management such as:

- Forecasts: They are obtained from data of potential stock for the different models in the different parcels where the simulation is carried out. They provide information about: the works to be done, the costs associated to them, the forecast profits for different years and for the end of turn or cycle of the technical and financial model developed.
- Situation reports: They are "snapshots" of the state and value of wood or biomass at a certain moment. They help determine the investment to be made over a certain period of time, the value of existing products in the parcels and the years and operations necessary to get profits from the proposed models.
- Investment and profitability analyses: They provide knowledge beforehand about the effort necessary to be made in forest activity in a certain parcel. Indicators such as IRR and NPV will be useful to calculate the expected profitability in each scenario or technical-economical model developed and the aspects that have more influence in such profitability.
- Improvement of training procedures (training/extension): The use of this application requires user training. Using this program improves the knowledge of the implications of forestry and forest management.
- Stock control: Users are able to consult the volume of wood that their parcels have at any given moment. This enables them to forecast their production or report losses in case of fire, wind or snow damage.
- Cost reduction: The availability of information entails a possible cost reduction. Two factors are key in this fact: first, the knowledge of average costs of the operations enables the management of the provider and contractor offers according to contrasted terms of reference; second, the knowledge of the optimum production models for each technical-economical models avoids unnecessary operations.
- Market transparency: The availability of accessible information has a direct influence in flux transparency and commercial relations that contribute to the clarification of a sector, which is traditionally considered to lack transparency. Knowing the real average cost of obtaining a product and its market value simplifies the buying and selling process and the accessibility to raw materials.
- Technical documentation and management system: Being an application based on a central server with all the security protocols, it is going to provide the users with a

space where they can make and store their operations with the guarantee of recovering and using them in the future.

- Help for administrative processes: A high number of activities related with forest activity productive processes are either publicly-funded or regulated by the administration. This administrative process comes with the need for technical documentation, which can be obtained with the format suitable for administrative use. In the future, electronic administration will be wide-spread so this application will be essential.

- Database with legislative information. The application has an associated database with legislative information (Lexplan module), with all the legislation regarding forest activity in its different phases. The user has access to all regulations and public funding that he or she can be eligible for according to the activity carried out in each particular moment

3.6 How does SaDDriade work?

The different sections below explain how SaDDriade works. We will start by the explanation of the programming language(s) used, the operative environment, the architecture goals and how it actually works.

3.6.1 Programming language

SaDDriade has been designed using more than one programming language. This was due to the complexity of the calculation processes, the web environment, the diversity of data sources, the need to have a GIS WEB tool and the different formats of exporting results. The languages used are: PHP (PHP Hypertext Preprocessor), Javascript, Mapscript and SQL (structured Query Language) All the components used in this application are open source components. They have been selected not only because of our agreement with the social philosophy and support of knowledge of the open source movement, but also for the financial advantages for both users and developers.

3.6.2 Operating environment

The main component of the Geographic Information System (In Spanish Sistema de Información Geográfica, SIG) integrated in the application is the open source platform Mapserver. This application was created in order to publish spatial information and to create interactive applications for maps. SaDDriade uses a combination of servers of relational databases of different kinds: on the one hand MySQL, and on the other hand PostgreSQL, with support for spatial data by means of the extension POSTGIS. Such services, together with a variety of files in shape format are the ones that feed the map server. The graphic interface was made in XHTML (eXtensible Hypertext Markup Language), a markup language whose specifications are developed by the World Wide Web Consortium (W3C).

3.6.3 Architecture

SaDDriade is a web application, so it can be used by accessing a web server by means of a client, typically a web browser. Through this client-server scheme, there is no need for a specific installation client side. It also makes it easier to install and maintain the application without having to distribute specific software to the clients.

The model-view-controller (MVC) design pattern was used to develop an internationalizable modular application which is easily extensible and has the capacity to access different database management systems. Three different frameworks were used, each of them with specific functions to optimize the general work of all the decision support system. **Codeigniter** framework is the base of the structure. Some of its advantages are: its speed, its excellent documentation, PHP compatibility backwards, small software and hardware requirements, native support for user-friendly URLs, drivers for a wide range of server database (MySQL, PostgreSQL, SQL server, SQLite, among others) extensible by helpers, plugins and libraries.

Codeigniter optimizes its performance eliminating non-essential distribution elements. Thus, its functionalities are more limited than those of other frameworks. In order to solve this problem, SaDDriade includes **Zend Framework**. Its main features are: low coupling among its components, wide unit test code coverage, multipurpose (capacity to generate PDFs, Access to LDAP, SOAP, Lucene, DOM, JSON, ACL system, email, authentication, automatic pagination, among others), use of design patterns, access to database by PDO and possibility to have an ORM (Object Relational Mapping), interoperatibility with the most important web services (Akismet, Amazon, AudioScrobbler, Delicious, Flickr, Nirvanix, Recaptcha, OpenID, Technorati, Twitter, Yahoo, Google, Youtube and Picassa) and high frequency of updates and new versions.

The third framework used is **Jquery,** a javascript library oriented to objects in the form of a resource collection that facilitates the reuse of the code and ensures the compatibility between different versions and types of browser. Special care was taken so as not to make an intrusive use, following the recommendations of the World Wide Web Consortium regarding accessibility. Within the general scheme of the application Jquery is responsible for the improvement of the user interface such as sortable tables and asynchronous communication with the server to increase the speed by the development of the XmlHttpRequest API (Application Programming Interface), a wide-spread technique known as AJAX (Asynchronous Javascript And XML).

SaDDriade GIS WEB structure revolves around Mapserver, which can process geographical information from different sources (WMS servers, WFS, shapes, databases, raster,...) and create very complex representations (depending on their configuration) which are shown in a client map. This configuration is normally done via GET parameters through a URL to a CGI script. It is an easy way to send requests to be analyzed by the server. A downside is that the possibilities of a dynamic answer using this method are limited and don't allow a complex application creation. To avoid this problem Mapserver has its own language, mapscript, which can access its API directly. In this way, programmers have all its library of functions to create interactive applications or RIA (Rich Internet Applications).

Mapserver itself can show the maps it generates or use a client to do so. SaDDriade uses a client. Maps are visualized using pmapper, a set of libraries made in PHP, jquery and mapscript, that offer more possibilities than those included in mapserver as a light client map. Pmapper has General Public License, so it can be used for free. It was chosen for its capacities, for being user-friendly and its integration with other program components. For instance, they share the jquery framework and the TCPDF library to create PDFs.

From the data obtained in the parcel shape and the zoning ecological criteria, pmapper displays a GIS client with capacity to add orthophotos of the SIGPAC project, parcel and municipality search, measurements, identification, zoom, annotation and map downloads as image or PDF. The user can choose a parcel, by clicking the "Identificar" button on the right side menu, and start the simulation process using the link offered by pmapper.

3.6.4 Work description

In SaDDriade, there are forest management models implemented for twelve different species (see table 1). In this way, 146 models have been parametrized in the forty areas in which Galicia has been divided. This has encompassed 13,108 tasks and subtasks, and the use of 160 different types of materials, machinery and so on.

Species	Number of models
Pinus pinaster	17
Pinus sylvestris	9
Pinus radiata	15
Pseudotsuga menziesii	8
Quercus robur	8
Quercus rubra	12
Quercus pyrenaica	6
Juglans regia	6
Populus sp. (hybrid)	8
Eucalyptus nitens	7
Castanea sativa	23
Castanea hybrid	27
TOTAL	146

Table 1. Species and number of models developed for each species.

Designed according to technical criteria:
- Possible ways of mechanization.
- Selection of the best available technique.
- Limit of appropriate densities.
- Programming activities on land and trees.
- Programming of intermediate harvests.

Financial criteria:
- Establishing technical shifts.
- Maximum rent shifts.
- Expense minimization.
- Benefit estimate.

Model choice

The first thing that the user must do is to choose which available SaDDriade module he/she would like to use: SAD Castanea, SAD Eucalyptus, SAD Pinus, SAD Populus and SAD Quercus.

Starting from the GIS-WEB, as stated above, the actual process starts once the user selects his/her parcel. Once the link shown is clicked, a window appears with basic data regarding the choice. Questions guide the user throughout the decision support process. By the location of a chosen parcel, a first filter of qualities and species has been set, so only those technical-economical models considered ecologically and financially viable are accessed. Models are classified by species and production destination to simplify the choice. The user must select a model among all the options to continue the calculations.

Fig. 2. SaDDriade initial screen.

Management variables

After the basic characteristics of the model are presented, some parameters that influence the task development and performance of the chosen model are shown: Characteristics of the place: slope, percentage of rocks, stone content, soil type; Initial state of vegetation: type, height, density and presence of stumps; Management alternatives: choice between base-root

Fig. 3. Characteristics of a selected model.

and pot planting, work line and plant and plantation protection; other management alternatives: characterization of protection and grubbing.

Options arise parallel to the simulation process and the user always has the option that the SaDDriade "remembers" his/her answers throughout the process just by clicking on the checkbox beside each answer. Twelve thousand possible combinations of profits, 13,000 yearly subtasks (that generate more than 4,300 partial products), 38 types of different machinery, 72 types of inputs and 47 different types of final outputs. They are all associated to prize, restriction, modelization and classification databases.

Results

Once the simulation is finished, a screen appears with the results in different formats.

- **Interactive support graphs.** They are interactive graphs that show the financial evolution of planting in time, broken down into expenses and income and cost distribution of labor and machinery.

- **Economic indicators. Total expenses:** result of the sum of all labor, machinery and input expenses. Total income: sum of the income obtained by product sale. Profits: Difference between income and expenses. NPV: Net Present Value. IRR: Internal Rate of Return calculated by the Newton- Raphson method.

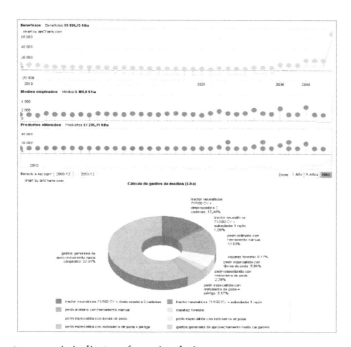

Fig. 4. Different economic indicators for a simulation.

- **Planting visualization with Google Earth**: The evolution of the planting of the parcel is shown by means of a Google Earth plugin (at this moment only available for Windows and Mac OS X 10.4 +). This enables a landscape analysis of the different decisions and operations carried out. It is a 3D representation of a planting on the square of the whole parcel or just a section, where the user can fly virtually into the trees, "walk" between them or make visualizations within a wider context.

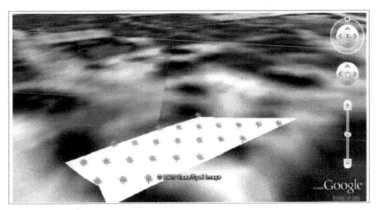

Fig. 5. Location of trees in a parcel and selected model.

Exportation

Exportation can be made in three formats, depending on the interest of the user: **kmz** (Compressed file containing geographical data and tridimensional models used to present the evolution of time in the planting. It can be opened with Google Earth), **pdf y xls.**

4. Conclusions¶

Decision Support Systems have proven to be useful in the different economic fields in which they have been developed because of their capacity of simulation and optimization. Forest DSSs have evolved over time thanks to IT, on the one hand, and due to the need to introduce higher social and environmental restrictions to forest management, on the other hand.

The development presented in this chapter includes the most advanced techniques in IT (web application and virtual reality simulation). It is also easy to use, which would allow a higher number of users to have access to it without much IT or forestry knowledge. Thus, it could, and it should, become a tool for forestry extension, which would make this type of management more sustainable and will give the possibility of increasing its level of technification.

New lines of development that will result into a better tool to the service of owners and forest managers are: the inclusion of optimization options and stochastic processes, the resolution by heuristic methods in which the risk of forest activities in different geographic environments will be taken into consideration and an improvement in the visualization options of virtual reality.

5. References

Acuña, S., Juristo, N., and Recio B., (1997). Knowledge-based system for generating administrative grant alternatives applying the IDEAL methodology. *Computers and Electronics in Agriculture* 18, 1-28. ISSN: 0168-1699

Aertsen, W., Kint, V., Van Orshoven, J., Özkan, K., and Muys, B., ((2010)). Comparison and ranking of different modelling techniques for prediction of site index in Mediterranean mountain forests. *Ecological Modelling* 221 ((2010)) 1119–1130.

Aitkenhead, M.J., and Aalders, I.H., ((2009)). Predicting land cover using GIS, Bayesian and evolutionary algorithm methods. *Journal of Environmental Management* 90, 236-250. ISSN: 0301-4797

Andersson, M., Dahlin, B., and Mossberg, M., ((2005)). The Forest Time Machine – a multi-purpose forest management decision-support system. *Computers and Electronics in Agriculture* 49, 114–128. ISSN: 0168-1699

Ascough II, J.C., Maier, H.R., Ravalico, J.K., and Strudley, M.W., ((2008)). Future research challenges for incorporation of uncertainty in environmental and ecological decision-making. *Ecological Modelling* 219, 383–399. ISSN: 0301-4797

Baskent, E.Z., Wightman, R.A., Jordan, G.A., and Zhai, Y., ((2001)). Object-oriented abstraction of contemporary forest management design. *Ecological Modelling* 143, 147–164. ISSN: 0301-4797

Bettinger, P., Lennette, M., Johnson,K.N., and Spies, T.A., ((2005)). A hierarchical spatial framework for forest landscape planning. *Ecological Modelling* 182, 25–48. ISSN: 0301-4797

Bonazountas, M., Kallidromitou, D., Kassomenos, P., and Passas, N., ((2007)). A decision support system for managing forest fire casualties *Journal of Environmental Management* 84, 412–418. ISSN: 0301-4797

Borchers, J.G., ((2005)). Accepting uncertainty, assessing risk: Decision quality in managing wildfire, forest resource values, and new technology. *Forest Ecology and Management* 211, 36–46. ISSN: 0378-1127

Boyland, M., Nelson, J., Bunnell, F.L., and D'Eon, R.G., ((2006)). An application of fuzzy set theory for seral-class constraints in forest planning models *Forest Ecology and Management* 223, 395–402. ISSN: 0378-1127

Bryan, B.A., and Crossman, N.D., (2008). Systematic regional planning for multiple objective natural resource management. *Journal of Environmental Management* 88, 1175–1189. ISSN: 0301-4797

Carlsson, M., Andersson, M., Dahlin,B., and Sallnas, O., 1998. Spatial patterns of habitat protection in areas with non-industrial private forestry – hypotheses and implications. *Forest Ecology and Management* 107, 203–211. ISSN: 0378-1127

Chertov, O., Komarov, A., Andrienko, G., Andrienko, N., and Gatalsky, P., (2002). Integrating forest simulation models and spatial–temporal interactive visualisation for decision making at landscape level. *Ecological Modelling* 148, 47–65. ISSN: 0301-4797

Crookston, N.L., and Dixon, G.E., (2005).The forest vegetation simulator: A review of its structure, content, and applications. *Computers and Electronics in Agriculture* 49, 60–80. ISSN: 0168-1699

Diaz-Balteiro, L., and Romero, C., (2004). Sustainability of forest management plans: a discrete goal programming approach *Journal of Environmental Management* 71, 351–359. ISSN: 0301-4797

Diaz-Balteiro, L., Romero, C., (2008). Making forestry decisions with multiple criteria: A review and an assessment. *Forest Ecology and Management* 255, 3222–3241. ISSN: 0378-1127

Ducey, M.J., Larson, B.C., 1999. A fuzzy set approach to the problem of sustainability. *Forest Ecology and Management* 115, 29–40. ISSN: 0378-1127

Ducheyne, E.I., De Wulf, R.R., De Baets, B., (2004). Single versus multiple objective genetic algorithms for solving the even-flow forest management problem. *Forest Ecology and Management* 201, 259–273. ISSN: 0378-1127

Falcão, A.O., dos Santos, M.P., , Borges, J.G., (2006). A real-time visualization tool for forest ecosystem management decision support. *Computers and Electronics in Agriculture* 53, 3–12. ISSN: 0168-1699

FEARMAGA, Monte Industria, Cluster de la Madera, feceg, (2009). Informe de resultadod de la industria forestal gallega (2008). http://monteindustria2.blogspot.com/search/label/informe 15-03-(2011).

Gärtner, S., Reynolds, K.M., Hessburg, P.F., Hummel, S., and Twery, M., (2008). Decision support for evaluating landscape departure and prioritizing forest management activities in a changing environment. *Forest Ecology and Management* 256, 1666–1676. ISSN: 0378-1127

Gustafson, E.J., and Rasmussen, L.V., (2002). Assessing the spatial implications of interactions among strategic forest management options using a Windows-based harvest simulator. *Computers and Electronics in Agriculture* 33, 179–196. ISSN: 0168-1699

Goldstein,M.I., Corson,M.S., Lacher Jr., T.E., and Grant, W.E., (2003). Managed forests and migratory bird populations: evaluating spatial configurations through simulation *Ecological Modelling* 162, 155–175. ISSN: 0301-4797

Graymore, M.L.M., Wallis, A.M., and Richards, A.J., (2009). An Index of Regional Sustainability: A GIS-based multiple criteria analysis decision support system for progressing sustainability *Ecological Complexity* 6, 453–462. ISSN: 1476-945X

Hackett, C., and Vanclay, J.K., 1998. Mobilizing expert knowledge of tree growth with the PLANTGRO and INFER systems. *Ecological Modelling* 106, 233–246. ISSN: 0301-4797

Heinimann, H.R.,(2010). A concept in adaptive ecosystem management—An engineering perspective. *Forest Ecology and Management* 259, 848–856. ISSN: 0378-1127

Huang, G.H., Sun, W., Nie, X., Qin, X., and Zhang, X., (2010). Development of a decision-support system for rural eco-environmental management in Yongxin County, Jiangxi Province, China. *Environmental Modelling and Software* 25, 24–42. ISSN: 1364-8152

Huth, A., Drechsler, M., and Köhler, P., (2005). Using multicriteria decision analysis and a forest growth model to assess impacts of tree harvesting in Dipterocarp lowland rain forests. *Forest Ecology and Management* 207, 215–232. ISSN: 0378-1127

Kaloudis, S., Tocatlidou, A., Lorentzos, N.A., Sideridis, A.B., and Karteris, M., (2005). Assessing Wildfire Destruction Danger: a Decision Support System Incorporating Uncertainty. *Ecological Modelling* 181, 25–38. ISSN: 0301-4797

Kangas, A.S., and Kangas, J., (2004). Probability, possibility and evidence: approaches to consider risk and uncertainty in forestry decision analysis. *Forest Policy and Economics* 6, 169–188. ISSN: 1389-9341

Kangas, J., and Kangas, A., (2005). Multiple criteria decision support in forest management—the approach, methods applied, and experiences gained. *Forest Ecology and Management* 207, 133–143. ISSN: 0378-1127

Kangas, J., and Leskinen, P. (2005). Modelling ecological expertise for forest planning calculations-rationale, examples, and pitfalls. *Journal of Environmental Management* 76, 125–133. ISSN: 0301-4797

Kangas, A., Laukkanen, S., and Kangas, J. (2006). Social choice theory and its applications in sustainable forest management—a review. *Forest Policy and Economics* 9, 77– 92. ISSN: 1389-9341

Kolström, M., and Lumatjärvi, J. 1999. Decision support system for studying effect of forest management on species richness in boreal forests. *Ecological Modelling* 119, 43–55. ISSN: 0301-4797

Kurttila, M., (2001). The spatial structure of forests in the optimization calculations of forest planning – a landscape ecological perspective. *Forest Ecology and Management* 142, 129-142. ISSN: 0378-1127

Kurz, W.A., Dymond, C.C., White, T.M., Stinson, G., Shaw, C.H., Rampley, G.J., Smyth, C., Simpson, B.N., Neilson, E.T., Trofymow, J.A., Metsaranta, J., and Apps, M.J., (2009). CBM-CFS3: A model of carbon-dynamics in forestry and land-use change implementing IPCC standards. *Ecological Modelling* 220, 480-504. ISSN: 0301-4797

Lasch, P., Badeck, F., Suckow, F., Lindner, M., and Mohr, P., (2005). Model-based analysis of management alternatives at stand and regional level in Brandenburg (Germany). *Forest Ecology and Management*, 207, 59–74. ISSN: 0378-1127

Leskinen, P., Kangas, J., and Pasanen, A.M., (2003). Assessing ecological values with dependent explanatory variables in multi-criteria forest ecosystem management. *Ecological Modelling* 170, 1–12. ISSN: 0301-4797

Li, H., Gartner, D.I., Mou, P., and Trettin, C.C., 2000. A landscape model (LEEMATH) to evaluate effects of management impacts on timber and wildlife habitat. *Computers and Electronics in Agriculture* 27, 263–292. ISSN: 0168-1699

Louw, J., and Scholes, M., (2002). Forest site classification and evaluation: a South African perspective. *Forest Ecology and Management*, 171, 153-168. ISSN: 0378-1127

MacMillan, D.C., and Marshall, K., (2004). Optimising capercailzie habitat in commercial forestry plantations. *Forest Ecology and Management*, 198, 351–365. ISSN: 0378-1127

Marey-Pérez, M.F., Crecente-Maseda, R., and Rodríguez-Vicente, V., (2006). Using GIS to measure changes in the temporal and spatial dynamics of forestland: experiences from north-west Spain. *Forestry*, 79 409–423. ISSN 1464-3626

Marey-Pérez, M.F., and Rodríguez-Vicente, V., (2008). Forest transition in Northern Spain: Local responses on large-scale programmes of field-afforestation. *Land Use Policy* 26, 139–156. ISSN: 0264-8377

Marey-Pérez, M.F., and Díaz-Varela, E.R., (2010). *El Sector Forestal. Plan Estratégico de la Provincia de Lugo.* Fundación Caixa Galicia y Diputación Provincial de Lugo. I.S.B.N.: 84-8192-664-7.

Martín-Fernández, S., and García-Abril, A., (2005). Optimisation of spatial allocation of forestry activities within a forest stand. *Computers and Electronics in Agriculture* 49, 159–174. ISSN: 0168-1699

Martins, H., and Borges, J.G., (2007). Addressing collaborative planning methods and tools in forest management. *Forest Ecology and Management*, 248, 107–118. ISSN: 0378-1127

McVicar, T.R., Li, L., Van Niel, T.G., Zhang, L., Li, R., Yang, Q., Zhang, X., Mu,X., Wen, Z., Liu, W., Zhao, Y., Liu, Z., and Gao, P., (2007). Developing a decision support tool for China's re-vegetation program: Simulating regional impacts of afforestation on average annual streamflow in the Loess Plateau. *Forest Ecology and Management* 251, 65–81. ISSN: 0378-1127

Matthews, K.B., Sibbald, A.R., and Craw, S., 1999. Implementation of a spatial decision support system for rural land use planning: integrating geographic information system and environmental models with search and optimisation algorithms. *Computers and Electronics in Agriculture* 23, 9–26. ISSN: 0168-1699

Meitner, M.J., Sheppard, S.R.J., Cavens, D., Gandy, R., Picard, P., Harshaw, H., and Harrison, D., (2005). The multiple roles of environmental data visualization in evaluating alternative forest management strategies. *Computers and Electronics in Agriculture* 49, 192–205. ISSN: 0168-1699

Mendoza, G.A., and Prabhu, R., (2005). Combining participatory modeling and multi-criteria analysis for community-based forest management. *Forest Ecology and Management* 207, 145–156. ISSN: 0378-1127

Mendoza, G.A., and Martins, H., (2006). Multi-criteria decision analysis in natural resource management: A critical review of methods and new modelling paradigms. *Forest Ecology and Management* 230, 1–22. ISSN: 0378-1127

Mette, T., Albrecht, A., Ammer, C., Biber, P., Kohnle, U., and Pretzsch, H., (2009). Evaluation of the forest growth simulator SILVA on dominant trees in mature mixed Silver fir-Norway spruce stands in South-West Germany. *Ecological Modelling* 220, 1670–1680. ISSN: 0301-4797

Mickovski, S.B., Stokes, A., and Van Beek, L.P.H., (2005). A decision support tool for windthrow hazard assessment and prevention. *Forest Ecology and Management* 216, 64–76. ISSN: 0378-1127

Mowrer, H.T., 2000. Uncertainty in natural resource decision support systems: sources, interpretation, and importance. *Computers and Electronics in Agriculture* 27, 139–154. ISSN: 0168-1699

Næsset, E., 1997. Geographical information systems in long-term forest management and planning with special reference to preservation of biological diversity: a review. *Forest Ecology and Management* 93, 121-136. ISSN: 0378-1127

Newton, P.F., (2003). Stand density management decision-support program for simulating multiple thinning regimes within black spruce plantations. *Computers and Electronics in Agriculture* 38, 45-53. ISSN: 0168-1699

Newton, P.F., (2009). Development of an integrated decision-support model for density management within jack pine stand-types. *Ecological Modelling* 220, 3301–3324. ISSN: 0301-4797

Nute, D., Rosenberg, G., Nath, S., Verma, B., Rauscher, H.M., Twery, M.J., and Grove, M., 2000. Goals and goal orientation in decision support systems for ecosystem management. *Computers and Electronics in Agriculture* 27, 355–375. ISSN: 0168-1699

Nute, D., Potter, W.D., Cheng,Z., Dass, M., Glende, A., Maierv, F., Routh, C., Uchiyama, H., Wang, J., Witzig, S., Twery, M., Knopp, P., Thomasma, S., and Rauscher, H.M., (2005). A method for integrating multiple components in a decision support system. *Computers and Electronics in Agriculture* 49, 44–59. ISSN: 0168-1699

Olofsson, E., and Blennow, K., (2005). Decision support for identifying spruce forest stand edges with high probability of wind damage. *Forest Ecology and Management* 207, 87–98. ISSN: 0378-1127

Porté, A., and Bartelink, H.H., (2002). Modelling mixed forest growth: a review of models for forest management. *Ecological Modelling* 150, 141–188. ISSN: 0301-4797

Potter, W.D., Liu, S., Deng, X., and Rauscher, H.M., 2000. Using DCOM to support interoperability in forest ecosystem management decision support systems. *Computers and Electronics in Agriculture* 27, 335–354. ISSN: 0168-1699

Rauscher, H.M., 1999. Ecosystem management decision support for federal forests in the United States: A review. *Forest Ecology and Management* 114, 173-197. ISSN: 0378-1127

Rauscher, H.M., Lloyd, F.T., Loftis, D.L., and Twery, M.J., 2000. A practical decision-analysis process for forest ecosystem management. *Computers and Electronics in Agriculture* 27, 195–226. ISSN: 0168-1699

Reynolds, K.M., (2005). Integrated decision support for sustainable forest management in the United States: Fact or fiction? *Computers and Electronics in Agriculture* 49, 6–23. ISSN: 0168-1699

Rodríguez-Vicente, V., and Marey-Pérez, M.F., (2010). Analysis of individual private forestry in northern Spain according to economic factors related to management. *Journal of Forest Economics* 16, 269–295

Salminen, H., Lehtonen, M., and Hynynen, J., (2005). Reusing legacy FORTRAN in the MOTTI growth and yield simulator. *Computers and Electronics in Agriculture* 49, 103–113. ISSN: 0168-1699

Seely, B., Nelson, J., Wells,R., Peter, B., Meitner, M., Anderson, A., Harshaw, H., Sheppard, S., Bunnell, F.L., Kimmins, H., and Harrison, D., (2004).The application of a hierarchical, decision-support system to evaluate multi-objective forest management strategies: a case study in northeastern British Columbia, Canada. *Forest Ecology and Management* 199, 283–305. ISSN: 0378-1127

Sheppard, S.R.J., and Meitner, M., (2005). Using multi-criteria analysis and visualisation for sustainable forest management planning with stakeholder groups. *Forest Ecology and Management* 207, 171–187. ISSN: 0378-1127

Schuster, E.G., Leefers, L.A., and Thompson, J.E., 1993. *A Guide to Computer-Based Analytical Tools for Implementing National Forest Plans. General Technical Report* INT-296. USDA Forest Service, Intermountain Research Station.

Snow, V.O., and Lovatt, S.J., (2008). A general planner for agro-ecosystem models. *Computers and Electronics in Agriculture* 60, 201–211. ISSN: 0168-1699

Stirn, L.Z., (2006). Integrating the fuzzy analytic hierarchy process with dynamic programming approach for determining the optimal forest management decisions. *Ecological Modelling* 194, 296-305. ISSN: 0301-4797

Store, R., and Jokimäki, J., (2003). A GIS-based multi-scale approach to habitat suitability modelling. *Ecological Modelling* 169, 1–15. ISSN: 0301-4797

Strange, N., Tarp, P., Helles, F., and Brodie, J.D., 1999. A four-stage approach to evaluate management alternatives in multiple-use forestry. *Forest Ecology and Management* 124, 79-91. ISSN: 0378-1127

Thompson, W.A., Vertinsky, I., Schreier, H., and Blackwell, B.A., 2000. Using forest fire hazard modelling in multiple use. forest management planning. *Forest Ecology and Management* 134, 163-176. ISSN: 0378-1127

Thomson,A.J., and Willoughby, I., (2004). A web-based expert system for advising on herbicide use in Great Britain. *Computers and Electronics in Agriculture* 42, 43–49. ISSN: 0168-1699

Thomson, A.J., Callan,B.E., and Dennis, J.J., (2007). A knowledge ecosystem perspective on development of web-based technologies in support of sustainable forestry. *Computers and Electronics in Agriculture* 59, 21–30. ISSN: 0168-1699

Torres-Rojo, J.M., and Sánchez Orois, S., (2005). A decision support system for optimizing the conversion of rotation forest stands to continuous cover forest stands. *Forest Ecology and Management* 207, 109–120. ISSN: 0378-1127

Twery, M.J., Rauscher, H.M., Bennett, D.J., Thomasma, S.A., Stout, S.L., Palmer, J.F., Hoffman, R.E., DeCalesta, D.S., Gustafson, E., Cleveland, H., Grove, J.M., Nute, D., Kim, G., and Kollasch, R.P., 2000. NED-1: integrated analyses for forest stewardship decisions. *Computers and Electronics in Agriculture* 27, 167–193. ISSN: 0168-1699

Twery, M.J., and Hornbeck, J.W., (2001). Incorporating water goals into forest management decisions at a local level. *Forest Ecology and Management* 143, 87-93. ISSN: 0378-1127

Twery, M.J., Knopp, P.D., Thomasma, S.A., Rauscher, H.M., Nute, D.E., Potter, W.D., Maier, F., Wang, J., Dass, M., Uchiyama, H., Glende,A., and Hoffman, R.E., (2005). NED-2: A decision support system for integrated forest ecosystem management. *Computers and Electronics in Agriculture* 49, 24–43. ISSN: 0168-1699

Vacik, H., and Lexer, M.J.., (2001). Application of a spatial decision support system in managing the protection forests of Vienna for sustained yield of water resources. *Forest Ecology and Management* 143, 65-76. ISSN: 0378-1127

Varma, V.K., Ferguson, I., and Wild, I., 2000. Decision support system for the sustainable forest management. *Forest Ecology and Management* 128, 49-55. ISSN: 0378-1127

Wang, J., Chen, J., Ju., W., and Li, M., (2010). IA-SDSS: A GIS-based land use decision support system with consideration of carbon sequestration- *Environmental Modelling and Software* 25, 539–553.

Wolfslehner, B., Vacik, H., and Lexer, M.J., (2005). Application of the analytic network process in multi-criteria analysis of sustainable forest management. *Forest Ecology and Management* 207, 157–170. ISSN: 0378-1127

Wolfslehner, B., and Vacik, H., (2008). Evaluating sustainable forest management strategies with the Analytic Network Process in a Pressure-State-Response framework *Journal of Environmental Management* 88, 1–10. ISSN: 0301-4797

Zeng, H., Pukkala, T., and Peltola, H., (2007). The use of heuristic optimization in risk management of wind damage in forest planning. *Forest Ecology and Management* 241, 189–199. ISSN: 0378-1127

9

The Effect of Harvesting on Mangrove Forest Structure and the Use of Matrix Modelling to Determine Sustainable Harvesting Practices in South Africa

Anusha Rajkaran and Janine B. Adams
Nelson Mandela Metropolitan University
South Africa

1. Introduction

Mangrove forests exist along a transitional boundary between land and sea. They represent a continuum of biotic communities between terrestrial and marine environments (Hogarth, 1999; Kathiresan and Bingham, 2001; Alongi, 2008). These forests are globally distributed between the subtropical and tropical latitudes, restricted by major ocean currents and the 20°C isotherm of seawater in winter (Hogarth, 1999; Alongi, 2009). On a global scale, temperature is an important limiting factor but on regional and local scales variations in rainfall, tides, waves and river flow have a substantial effect on distribution and biomass of mangrove forests (Alongi, 2009). Erosion and depositional rates are also important as these affect the physical habitat that mangroves occupy. Generally the habitat of mangroves begins at mean sea level and extends to the spring high tide mark i.e. they exist in tidal areas (Hogarth, 1999; Spalding et al., 2010) while in South Africa mangroves are confined to estuaries that either may be permanently open to the sea or have an intermitted connection to the sea (Rajkaran, 2011). Estuaries are defined as "a partially enclosed body of water which is either permanently or periodically open to the sea and within which there is a measurable variation of salinity due to the mixture of seawater with freshwater derived from land drainage" (Day, 1980) as being; river mouths, estuarine bays, permanently open estuaries, temporarily open closed and estuarine lakes. There are five types of estuaries and these are defined by Whitfield (1992). The ecosystem services provided by mangroves include; shoreline protection from sea storms and excessive wave energy, nursery and areas of refugia for faunal populations (BOX 1), input of organic carbon into the food webs and filtration of silt and other compounds from the water column thereby protecting other nearshore ecosystems such as coastal reefs (Gilbert & Janssen, 1998; Fondo & Martens, 1998; Laegdsgaard & Johnson, 2001; Mumby et al., 2003). Mangrove forests are known to have survived for approximately 65 million years and therefore are resilient to large scale disturbances (Alongi, 2009). Key mangrove features that have assisted in their resilient nature include; the presence of a large reservoir of below-ground nutrients so that if a disturbance takes place the remaining nutrients will assist with the re-establishment of new seedlings to replace those that have been lost encouraging re-population of the disturbed area. Rapid biotic turnover has been recorded in mangrove forests and is facilitated by rapid rates of nutrient flux and microbial decomposition. Internal recovery after a disturbance is accelerated by complex and efficient biotic controls such as nutrient-use efficiency (Alongi, 2008, 2009). Frequent, small scale disturbances such as harvesting disrupts the flow of nutrients from the living biomass to the sediment environment via the roots, it also facilitates changes to the microenvironment which will reduce the capacity of the mangrove forests to recover.

BOX 1.

Faunal diversity in mangrove forests is high including organisms from sponges to elasmobranchs and bony fish as well as bird species such as the Mangrove Kingfisher (Nagelkerken et al., 2008). Crabs are the most abundant macrofauna (numbers and biomass) in mangrove forests (Smith et al., 1991). They consume or hide 30 to 80 % of leaves, propagules and other litter on the floor of mangrove forests (Dahdouh-Guebas et al., 1997; Machiwa & Hallberg, 2002; Skov et al., 2002). Crabs enhance degradation of leaves and make the leaves available to meiofauna (Dahdouh-Guebas et al., 1999). The diversity of crabs found in a mangrove forest may vary. At Mngazana Estuary the following species were found *Neosarmatium meinerti* de Man, *Sesarma eulimene* de Man, *Sesarma catenata* Ortmann, *Uca lacteal annulipes* H. Milne Edwards, *Uca chlorophthalmus chlorophthalmus* (H. Milne Edwards), as well as *Parasesarma leptosome* (Hilgendorf) (Plate 1). The latter is a tree climbing crab that spends most of its life in the mangrove trees and is therefore totally dependent on mangrove forests for their existence (Emmerson et al., 2003; Emmerson & Ndenze, 2007). More recently the species *Perisesarma samawati* Gillikin & Schubart, which was only described to occur in East Africa was spotted at Mngazana Estuary in South Africa in 2011 for the first time (Plate 2).

Plate 1 and 2: Images of crab species only associated with mangrove forests. Photos taken by Anusha Rajkaran

---

2. Mangrove forests: Utilization and destruction

In 2003, the global estimate of mangrove forest cover was 14 650 000 ha and accounted for approximately 0.7% of the total global area of tropical forests (Wilkie & Fortuna, 2003; Giri et al., 2011). Each hectare is valued at between 200 000 – 900 000 USD (Wilkie & Fortuna, 2003; Giri et al., 2011). Human disturbances has resulted in more than 50% of the world's mangrove forests being destroyed (Spalding et al., 2010). This huge loss of mangrove forests globally, has been attributed to urban development, aquaculture, mining along coastal zones and overexploitation of fauna and flora of mangrove forests (Walters, 2005; Walter et al., 2008; Kairo et al., 2008; Alongi, 2009). The connection between coastal developments, water level fluctuations and mangrove loss or transformation has been recorded by a number of authors in South Africa and other parts of the world (Moll et al., 1971; Begg, 1984; Bruton, 1980;

Dahdouh-Guebas et al., 2005) (BOX 2). Worldwide, mangrove forests are harvested for a variety of purposes. The products are particularly important to subsistence economies, providing firewood, building supplies and other wood products (Bandaranyake, 1998; Ewel et al., 1998; Cole et al., 1999; Kairo et al., 2002; Dahdouh-Guebas et al., 2004, Walters et al., 2008). The subsequent effects on the ecosystem ranges from loss of habitat for fauna such as arboreal crabs (Emmerson and Ndenze, 2007), decreases in organic carbon export to the food webs and nearshore environments (Rajkaran & Adams, 2007), coastal erosion (Thampanya et al., 2006) and in the long term, loss of nursery functions (Laegdsgaard & Johnson, 2001).

BOX 2

Freshwater abstraction and poor bridge design has caused the mouths of some South African estuaries with mangroves to close to the sea more frequently, leading to long term inundation of roots and subsequent death of the mangroves (Breen & Hill, 1969; Bruton, 1980; Begg, 1984). Rising water levels have been one of the main factors that have lead to localised mangrove disturbances and mortalities in Kosi Bay (1965-1966) and Mgobezeleni Estuary (74 km south of Kosi Bay) (Bruton, 1980). Past data shows that 78% of the 1084 trees died in the Mgobezeleni Estuary due to submergence of the root structures when the water level rose for an extended period of time. This was a result of water being impounded behind a bridge constructed in 1971. Dead mangrove trees ranged from 40 cm to 15 m in height showing that all height classes are susceptible to death due to water level increases. The living mangrove stand became infested by the mangrove fern. In 2007, 77 *Brugueira gymnorrhiza* trees were still living, these have all since died (2011). The water level was ~ 30 cm of water above the sediment. Less than five seedlings were seen in areas where the sediment was not submerged. This estuary is a prime example of how poor coastal planning and developments can have a negative effect on surrounding coastal habitats such as mangrove forests.

Plate 3 and 4: Images taken at Mgobezeleni Estuary in 2007 by Dr. Ricky Taylor showing the submergence of the root structures of the *Bruguiera* trees and the extent of the mangrove fern.

2.1 Effect of harvesting on mangrove forests

Gaps created during the harvesting of either individual or groups of trees provide opportunities for seedling recruitment and growth (Rabinowitz 1978; Ewel et al., 1998; Sherman et al., 2000). The size class structure of mangrove forests in localities that experience harvesting show under-representation in large size classes, which is the result of selective harvesting (Saifullah et al., 1994; Walters 2005). Because mangrove wood is used for building, the size of the mangrove poles determines the role they play in the built structure. A comparison of height classes of the non-harvested and harvested sites in the Mngazana Estuary (31°42'S, 29°25' E) in South Africa showed that the height class 2.3 – 3.3 m was dominant in non-harvested sites while in harvested sites smaller trees were dominant. All the harvested poles were approximately 3 m (Rajkaran & Adams, 2010). Traynor & Hill (2008) interviewed harvesters with regard to harvesting preferences at Mngazana Estuary; they stated that any tree greater than 2 m in height with a desired diameter at breast height (DBH) would be harvested. They also stated that the required length of the wall poles used for building homesteads was 3 m for wall poles while roof poles were usually 4 m. This explained the differences found for mangrove height between harvested and non-harvested sites. Traynor & Hill (2008) recorded that the preferred species for building was *Rhizophora mucronata* (41% of participants preferred this species) and *Bruguiera gymnorrhiza* (21%) while *Avicennia marina* was used for firewood.

3. The use of matrix modelling to determine sustainable harvesting practices

With the use of population models one can predict the quantitative changes in population structure and thus add value to any management plan established for a particular mangrove forest. Mathematical models are popular conservation and management tools used to predict changes to plant and animal populations that are at risk due to activities such as harvesting (Raimondo & Donaldson, 2003; López-Hoffman et al., 2006; Owen-Smith, 2007; Ajonina, 2008). Matrix models are age or stage structured models used in cases when harvesting of particular size classes is the main risk. One takes into account the probability of an individual plant moving from one size class to the next i.e. transition probabilities as well as the possibility of the individuals persisting in the size class or dying (Caswell, 2001; Porte & Bartelink, 2002; Boyce et al., 2006; Owen-Smith, 2007; Caswell, 2009). In the case of plants, the model usually uses plant size (height or DBH) as the basis for the model. Model parameters include recruitment (the portion of propagules that is produced by a specific size class that is added to Size Class 1), mortality (M), transition rates (T) and persistence rates (P) for each size class, these are known as the vital rates (Caswell, 2001; Porte & Bartelink, 2002; Owen-Smith, 2007) (Figure 1).

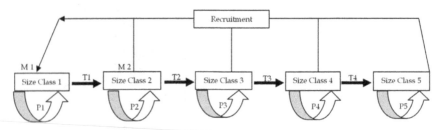

Fig. 1. The layout of the matrix model illustrating the vital rates mortality (M), transition rates (T) and persistence rates (P) for each size class.

The objective of this study was to develop a matrix model to determine the effect of different harvesting intensity scenarios, on the population structure of three mangrove species: *Avicennia marina*, *Bruguiera gymnorrhiza* and *Rhizophora mucronata*. The model results were compared to the observed population structure measured in the field at the end of the study in 2009 to determine the accuracy of the model and used to determine the most sensitive size classes to changes in vital rates within the population. Some data are presented here but more detailed results can be found in Rajkaran (2011).

3.1 Model development and accuracy
Nine sites at Mngazana Estuary were studied to collect data for the population model. This estuary is located in the Eastern Cape Province of South Africa, (Figure 2). In each site the following information was recorded, number of saplings (no hypocotyl less than 1 m), number of adults (over 1 m), the height of saplings and DBH and height of adults were measured. Subsequent measurements took place in November 2005, June 2006, November 2006, June 2007, November 2007, November 2008 and November 2009.

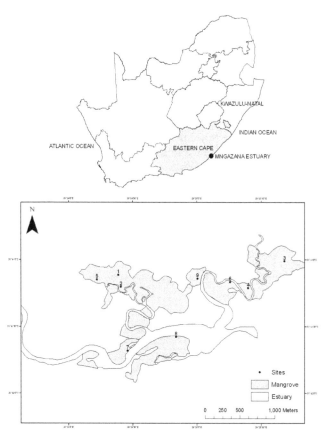

Fig. 2. The location of Mngazana Estuary in the Eastern Cape of the Republic of South Africa and the location of Sites 1-9 where growth was monitored from 2005-2009.

The population of each species, as calculated from nine sites around the estuary, was summarised and divided into a number of size classes based on mangrove height (Table 1). **Transition rates** were determined by counting the number of individuals in each size class over a period of five years (2005-2009). The **persistence rate** was the percentage of individuals that were in the same size class between two successive years (2005 compared to 2006). The transition rate was the percentage of individuals that were still alive but were now in the next successive size class therefore they had grown taller. Mortality rates were determined for the first two size class i.e. <50 cm and 50.5-150 cm height. The natural mortality of the other size classes could not be determined as none of the taller trees died unless they were harvested by the local community. In the model, natural mortality was included within the persistence rate i.e. the persistence rate was lowered by the appropriate percentage determined for each species based on the five year dataset. On two sampling trips (November 2005 and June 2006) the number of propagules on each tree was counted and the height of the tree was recorded. These data were used to determine the **fecundity** of each size class and were used as input on the proportion of propagules added by each size class to the total number of propagules.

Natural recruitment which was the number of new seedlings (hypocotyls present - <50 cm) added to the population was calculated for the five year period. Not all propagules that are produced establish themselves due to crab predation and removal by tidal movement. The number of individuals in each size class was converted from trees. m^{-2} (calculated from site data) to trees.ha^{-1}. The number of individuals that an area is able to support (carrying capacity) was assumed to be the total number of individuals in the population. The model was formulated to be density dependant, therefore the greater the number of individuals in the total population the stronger the effect of competition on the smaller individuals resulting in a lower survival rate. The time span for each population model was determined by how long the population size would take to stabilise. N_t is the size of the population at the start of the study. N_{t+1} is the sum of all the size classes calculated for each year after the start of the study (t+1). The ratio between N_{t+1} / N_t is the finite rate of increase and summarises the dynamics of a population. This ratio is symbolised by lambda (λ-the dominant eigenvalue of the matrix). When $\lambda=1$ then the population is in balance and remains stable ($N_{t+1} = N_t$), if $\lambda>1$ the population is increasing ($N_{t+1} > N_t$) and if $\lambda<1$ then the population is decreasing ($N_{t+1} < N_t$) (Slivertown & Charlesworth, 2001; Rockwood, 2006). Initial model results were compared to the observed population structure measured in the field at the end of the study in 2009 to determine the accuracy of the model.

3.2 Harvesting intensity scenarios

Harvesting scenarios represented a static harvesting rate of 1, 5, 10, 15, 20 and 100% of individuals for the three different species present at Mngazana Estuary. To determine the effect of harvesting on the total population (N) as well as different size classes a number of harvesting scenarios were added to the model. Population monitoring showed that harvesting of trees taller than 250 cm was common, therefore the model assumed that a percentage of Size Class 4 (250-350 cm) and 5 (>351 cm) would be harvested each year. The following harvesting intensities were used; 1, 5, 10, 15, 20 and 100% of a particular size class.ha^{-1}.year^{-1}. These scenarios would show how much of the population could be harvested and what the limit was for harvesting. The scenarios also showed how each size class changed in abundance in response to the different harvesting intensities.

3.3 Results

The *Avicennia marina* trees at Mngazana Estuary are either completely harvested or portions of the tree are cut for firewood. The assumptions for this model were 1) a tree, or portion of a tree, used for firewood is taken as a completely harvested tree and 2) that harvesting only affects the tallest trees in the forest (S5). The second assumption was based on field observations from Mngazana and Mhlathuze estuaries, where the tallest trees were the ones that were targeted. A hundred percent harvesting of individuals in the tallest size class decreased the total population to below 10 000 trees.ha^{-1} (Figure 3) and λ to 0.994 (Table 2). Restricting harvesting to just one size class that has reached reproductive maturity will ensure that other trees will still be present to produce propagules and subsequently seedlings. For this reason λ values as shown in Table 2 for *Avicennia* remain just below 1 for all harvesting scenarios. The number of individuals in Size Class 2 under 0% harvesting stabilised at less than 10 000 per ha (Figure 4). This decreased when the harvesting intensity increased as did the number of individuals in all size classes. To ensure more than 5 000 individuals were present in Size Class 1, which represents the main class for natural regeneration, harvesting must not exceed 20% of the trees taller than 350 cm per year. This is equivalent to 238 ± 4.5 harvested trees.ha^{-1}.yr^{-1}.

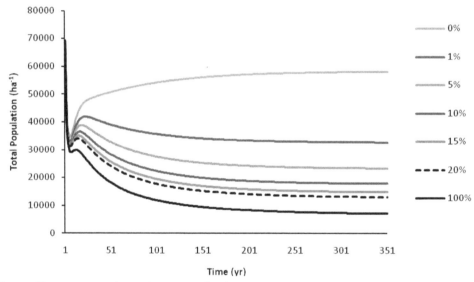

Fig. 3. Changes in total population size for the species *Avicennia marina* over time in response to different harvesting scenarios.

The assumption was that harvesting of two size classes would take place at Mngazana Estuary for *Bruguiera*. All trees greater than 251 cm would be removed. Harvesting of this species had a dramatic effect on the total population size. The total population of this species decreased by 63% when harvesting intensity was set at 1%. This allowed the population to stabilise at 15 000 trees.ha^{-1} (Figure 5). A further scenario was run using a harvesting intensity of 2%, this reduced the total population to approximately 5 000 trees.ha^{-1}. The mean λ for this species dropped from 0.999 to 0.834 at 100% harvesting intensity showing that the population was decreasing and natural regeneration was not taking place (Table 3).

Species	Size class (Height)	S1 <50 cm	S2 50-150 cm	S3 151-250 cm	S4 251-350 cm	S5 >351 cm
Avicennia marina	$N_{(t0)}$ (per ha^{-1})	16 786	40 536	8 036	2 500	1 339
	T	0.2	0.1	0.1	0.1	0
	P	0.6	0.8	0.9	0.9	0.9
	F	0	0	0	0.5	0.5
	MR (%)	21.0 ± 6.8	6.9 ± 2.0	ND	ND	ND
Bruguiera gymnorrhiza	$N_{(t0)}$ (per ha^{-1})	12 831	10 703	2 109	2 188	2 266
	TR	0.08	0.08	0.12	0.02	0
	PR	0.79	0.8	0.88	0.98	0.9
	F	0	0.16	0.16	0.33	0.33
	MR (%)	12.2 ± 4.6	7.2 ± 7.6	ND	ND	ND
Rhizophora mucronata	$N_{(t0)}$ (per ha^{-1})	11 979	43 750	10 104	8 125	2 917
	TR	0.3	0.03	0.1	0.03	0.1
	PR	0.6	0.88	0.9	0.97	0.9
	F	0	0.16	0.16	0.33	0.33
	MR (%)	15.6 ± 3.6	8.5 ± 2.3	ND	ND	ND

Table 1. Summary of data for each species and size class (S1-S5) used to populate the matrix models. (Transition rates (T) and persistence rates (P), fecundity rate (F), mortality rate (MR)).

Harvesting intensity	Total Population (N)	Size class (Height (cm))				
		0-49	50-150	151-250	251-350	>350
0%	1.000	1.001	0.997	1.001	1.004	1.006
1%	0.998	0.999	0.996	0.999	1.002	1.004
5%	0.997	0.998	0.995	0.998	1.002	1.002
10%	0.996	0.997	0.994	0.998	1.001	1.001
15%	0.996	0.997	0.994	0.997	1.001	0.999
20%	0.996	0.996	0.994	0.997	1.000	0.998
100%	0.994	0.995	0.992	0.996	0.999	0.999

Table 2. Mean λ values for *Avicenna marina* under different harvesting scenarios after 350 years.

Harvesting intensities of 15% and 100% were omitted from the graphs as the curves were similar to the 20% harvesting intensity and were not visible. Harvesting 1% of the adult trees maintained the density of size class 1 to < 5 000 individuals.ha⁻¹ (Figure 6).

The same assumption regarding harvesting was used for *Rhizophora mucronata* that harvesting of two size classes would take place at Mngazana Estuary. Documented data showed that the average length for harvested poles was 3.4 m. Harvesting scenarios in the model were restricted to the last two size classes (>251 cm). Total population size decreased from ~ 80 000 to 28 000 individuals.ha⁻¹ when harvesting intensity was 1%, this represented a 65 % reduction (Figure 7). λ values decreased to less than 1.000 showing that the population was decreasing as a result of the harvesting (Table 4). Harvesting intensity greater than 15% decreased the density of Size class 1 to ~3 500 individuals.ha⁻¹ (Figure 8). Harvesting between 5-10% of trees per year would amount to 183 – 283 harvested trees.ha⁻¹.yr⁻¹.

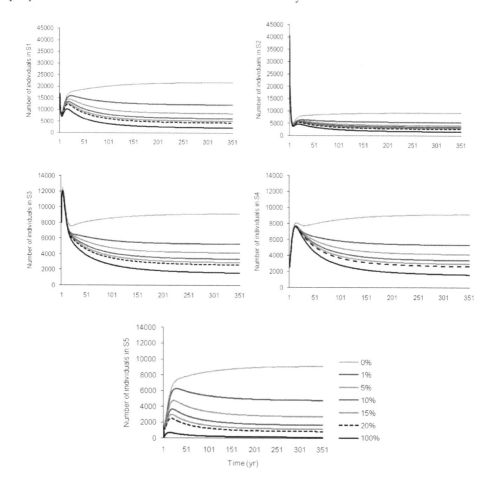

Fig. 4. The impact of harvesting on the number of individuals.ha⁻¹ in each size class of the *Avicennia marina* population over time. (Y-axis was not standardised for all graphs so that curves would be visible)

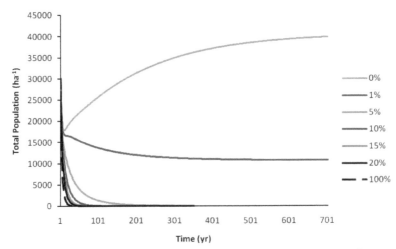

Fig. 5. Changes in total population number for the species *Bruguiera gymnorrhiza* over time in response to different harvesting scenarios.

Harvesting intensity	Total Population (N)	Size class (Height (cm))				
		0-49	50-150	151-250	251-350	>350
0%	1.000	1.001	0.999	1.000	1.002	1.000
1%	0.999	0.999	1.000	1.000	0.999	0.999
5%	0.980	0.980	0.979	0.981	0.983	0.977
10%	0.949	0.949	0.948	0.950	0.951	0.946
15%	0.922	0.922	0.922	0.924	0.924	0.918
20%	0.901	0.901	0.901	0.902	0.902	0.896
100%	0.834	0.832	0.835	0.834	0.832	0.822

Table 3. Mean λ values for *Bruguiera gymnorrhiza* for different harvesting scenarios after 701 years, the number of years required for the population to reach equilibrium was greater than for the other two species.

3.4 Discussion

Small scale disturbances such as harvesting, depending on the timing, frequency and intensity, which result in the loss of some of the mangrove population, may lead to natural regeneration if there are existing seedlings, saplings and mother trees (standard) around the disturbed area, if there is potential for water-borne propagules to travel to the area via tidal flow and if the propagules from disturbed trees are still present (FAO, 1994). A "standard" is defined as a seed bearing tree that can withstand exposure to strong winds and light and, in fringe areas, high tidal action (FAO, 1994). Regeneration will be restricted if the number of standards is reduced, if dead trees and branches reduce the light on the forest floor, if damage occurs to surrounding seedlings/saplings due to trampling and if a substantial change in soil conditions occurs (FAO, 1994; Harun-or-Rahsid et al., 2009).

Clarke et al., (2001) noted that the lack of diaspore dormancy in most mangrove species translates into a small or non-existent seed bank. The lack of a persistent soil seed bank of true mangrove species decreases the probability of a full recovery by mangrove populations after large scale disturbances and increases the chances of invasions of mangrove-associate species (Dahdouh-Guebas et al., 2005; Harun-or-Rahsid et al., 2009). Populations are reliant on regular cohorts of diaspores for regeneration so their continuous production by adults is vital. Rajkaran & Adams (2007) recorded movement of propagules out of the creeks and main channel of Mngazana Estuary, dispersed propagules were found on the adjacent beach near the mouth of the estuary. At Mngazana Estuary the presence of propagules on the forest floor is dependent on that produced by the adults in that specific area and not on the propagules brought in by tides. So at this estuary the continuous production by adults remaining after harvesting is vital for natural regeneration.

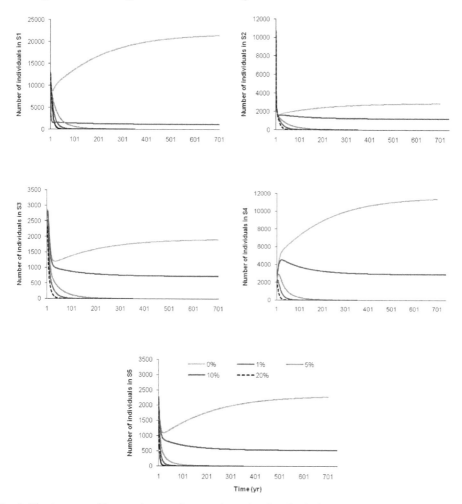

Fig. 6. The impact of harvesting on the number of individuals.ha^{-1} in each size class of the *Bruguiera gymnorrhiza* population over time. (Y-axis was not standardised for all graphs so that curves would be visible).

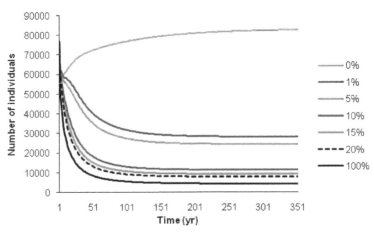

Fig. 7. Changes in total population size for the species *Rhizophora mucronata* over time in response to different harvesting scenarios.

Harvesting intensity	Total Population (N)	Size class (Height (cm))				
		0-49	50-150	151-250	251-350	>350
0%	1.000	1.003	0.998	0.999	1.003	1.002
1%	0.997	1.000	0.996	0.997	0.997	1.000
5%	0.997	0.999	0.996	0.996	0.997	0.999
10%	0.995	0.997	0.994	0.995	0.993	0.994
15%	0.994	0.996	0.994	0.994	0.992	0.993
20%	0.994	0.996	0.994	0.994	0.992	0.991
100%	0.992	0.994	0.992	0.992	1.001	0.974

Table 4. Mean λ values for *Rhizophora muronata* for different harvesting scenarios after 350 years.

Size classes in this study were based on height as previous studies have shown that harvesters targeted specific heights within the population (Rajkaran & Adams, 2009; Traynor & Hill, 2008). A density dependent model was used to simulate population structure and growth over time and the results conformed well to the logistical equation. The average λ value for each species in the absence of harvesting scenarios was 1.000, which shows that the populations are not increasing under the current harvesting rates for each size class. This may be a consequence of the continuous past harvesting in the Mngazana mangrove forest that has influenced vital rates. This was not taken into account in this model. López -Hoffman et al. (2006) recorded λ values of 1.050 when no harvesting was taking place. Vital rates for *Rhizophora mucronata* were comparable to those measured by López -Hoffman et al., (2006). Persistent rates ranged from 0.909 to 0.983, while transition rates ranged from 0.026-0.034 for adult size classes in that study, which is similar to the current study for this species. Similar studies for *Bruguiera gymnorrhiza* were not found. Clarke, (1995) used a matrix model to predict the population dynamics of *Avicennia marina* in New Zealand. Persistence rates for seedlings were 0.825, saplings - 0.909, young tree -

0.963 and older trees 0.999, while transition rates were 0.010, 0.073, 0.008, 0.012, 0.000 respectively. Sizes of each life stage were not stated in the study. The persistence rate in this study for *Avicennia* seedlings was much lower at 0.6 and transition was higher at 0.2, while all other rates were comparable with other studies. This implies that *A. marina* seedlings in South Africa grow faster and more seedlings survive to the next population size class within one year but the overall survival of the seedlings is similar between the two studies. Faster growth rates are dependant on site specific environmental conditions such as sediment characteristics and interspecific competition (Rajkaran, 2011).

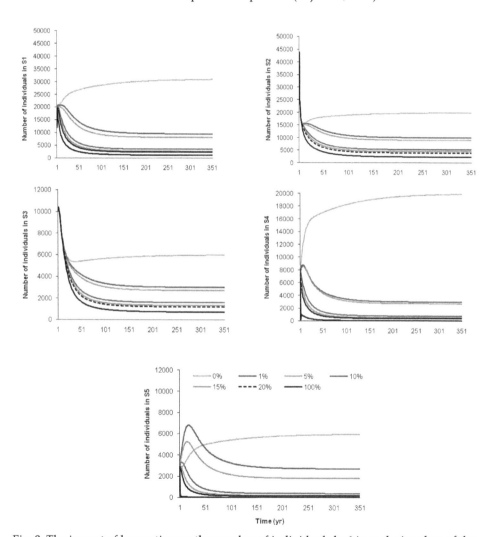

Fig. 8. The impact of harvesting on the number of individuals.ha⁻¹ in each size class of the *Rhizophora mucronata* population over time. (Y-axis was not standardised for all graphs so that curves would be visible for S5 and S3)

All harvesting scenarios decreased λ to less than 1.000, showing that the populations were decreasing in size. A sustainable harvesting rate would be one where λ is greater than 1. This would indicate that harvesting would be increasing the population growth by increasing space and decreasing competition between individuals. A λ value of 1.000 would mean that the population is unchanging (López -Hoffman et al., 2006) and disturbance would be detrimental to the population. FAO (1994) have set minimum limits for the number of "mangrove" seedlings that must be present to facilitate natural regeneration once adults have been removed from the population. The harvesting intensity that leads to a seedling density of less than 5000.ha^{-1} were 100% intensity for *Avicennia marina* all intensities greater than 1% for *Bruguiera gymnorrhiza* and 15, 20 and 100% for *Rhizophora mucronata*. The limits of harvesting in the Mngazana mangrove forest should not approach these levels. López -Hoffman et al., (2006) set sustainable harvesting in the Rio Limón mangrove forests of Lake Maracaibo in Venezuela at 7.7% per year for *Rhizophora mangle*, the current study has set harvesting limits at 5% per year for *Rhizophora mucronata* and *Avicennia marina*. Harvesting of *Bruguiera gymnorrhiza* should be stopped as the density of this species is lower than the other two species. Preferably there should be no harvesting of this species. Harvesting intensity must ensure that seedling density is maintained within acceptable limits as set out in the published literature (FAO, 1994; Bosire *et al.* 2008; Ashton & Macintosh, 2002). A density of 2 500 – 3 200 seedlings ha^{-1} has been suggested as a minimum number required for natural regeneration to take place after a disturbance (FAO, 1994; Bosire et al. 2008). Ashton and Macintosh (2002) recommended 5 000-10 000 seedlings ha^{-1} for adequate regeneration in a cleared area in the Matang Mangrove forest in Peninsular Malaysia. Density of individuals of the three species were measured at Mngazana Estuary in 2005 and were found to be 17 000, 13 000 and 12 000 seedlings.ha^{-1} for *Avicennia, Bruguiera* and *Rhizophora* respectively. To set the minimum number of seedlings to 5 000 individuals.ha^{-1} would mean that this size class would be more than half the original density. Increasing the limit to 10 000 seedlings.ha^{-1} would be more acceptable at the Mngazana Estuary for all species. The harvesting limits for each species will be different but managers must ensure that the seedling densities are maintained.

Mangrove management regimes may also suggest different densities for standards, i.e. the reproductively active trees producing propagules; these range from 7 (Malaysia) to 20.ha^{-1} (Phillipines) (Choudhury, 1997). This depends on the species; FAO (1994) suggested 12 standards.ha^{-1} for the genus *Rhizophora*. These levels are recommended for forests where clear-felling takes place in tropical countries where growth rates are high. Clear felling should be avoided in the Mngazana mangrove forest as this will significantly change sediment characteristics. Sediment conditions are significantly affected by changes in vegetation cover and plant density in a mangrove forest (Rajkaran and Adams, 2010). Mangrove forests are made up of species that are able to attain slow growth under a wide variety of conditions (Krauss et al., 2008) but Rajkaran and Adams (under review) recorded that growth and mortality of different size classes within a population were related to certain sediment parameters i.e. seedling growth was negatively related to high sediment pH (*Rhizophora* upper limit for pH in this study was 7.1) while seedling mortality for *Bruguiera* was negatively affected by an increase in sediment moisture.

A harvesting intensity of 5 % would maintain the number of individuals for *Rhizophora mucronata* at greater than 3 000.ha^{-1} in Size Class 3 and Size Class 4 while Size Class 5 would be reduced to approximately 2 000 individuals.ha^{-1}. Traynor & Hill (2008) estimated the annual demand for mangroves at 18 400 stems.yr^{-1} at Mngazana. These were mainly used by the local communities to build homesteads. The suggested harvesting intensity of between 5 and 10% per year would provide this required number of stems and indeed yield more harvested stems than those required at the time of the 2008 study. A more detailed study about the increase in the demand over time due to increases in the human population is required, but in the meanwhile an alternative wood resource must also be provided to the communities to replace the mangroves. The full effects of harvesting have not been measured in this study because, for example, the effects of trampling on seedling survival and its influence on population growth and structure were not addressed. Recruitment was extremely low in this study which may have been the influence of physical disturbance from harvesters. Other management recommendations include reducing harvesting within the 10 - 20 m strip from the estuary channel. The purpose would be to sustain trees that form a barrier between the energy of the water flowing in on a high tide and the young seedlings.

4. Management of mangrove systems in South Africa

The management of ecosystems calls for the interaction between researchers and society to ensure that environmental and socio-economic issues are integrated with government policies. For this to take place a number of conceptual frameworks exist as tools for communication between researchers and end users of environmental information such as government departments (Maxim et al., 2009). The Drivers-Pressures-Status-Impact-Response (DPSIR) framework focuses on the connecting relationships between the **Driving** forces that are usually societal and economic developments that place the environment under **Pressure** which alters the **State** of the environment, and **Impacts** on the ecosystems. The **Response** from society is usually in the form of regulatory laws or rehabilitation plans depending on the situation (Bidone & Lacerda, 2004; Maxim et al., 2009; Omann et al., 2009; Atkins et al., 2011). The DPSIR framework allows managers and scientists to highlight issues that must be prioritised with regard to management of natural systems. The DPSIR framework was applied to the results from this research and identifies the issues associated with the management of mangroves in South Africa (Figure 9).

Overall interventions for the conservation of mangroves in South Africa include directly protecting pristine mangroves, protecting the hydrological regimes supporting these ecosystems (particularly freshwater quantities flowing into the estuaries-which would be dependent on the base-flows required to maintain mouth conditions in the optimal state), promoting natural regeneration for self renewal, enforcing mangrove buffer zones and the continued capacity development and education of those communities that use the forests (Macintosh & Ashton, 2004). Mangrove buffer zones provide protection to any habitat or human areas behind them. Vietnam maintains a 100 m – 500 m wide belt of mangroves to protect the Mekong Delta coastline against storm and flood protection, while the Philippines maintain a 20 m wide zone for protection of shorelines (Macintosh & Ashton, 2004). All mangroves in South Africa are found within estuarine ecosystems so their capacity to protect the coastline is limited. However in many cases coastal developments have occurred along the banks of estuaries behind mangrove and salt marsh communities. In these cases it is recommended that a mangrove buffer zone of 25 m be maintained and in the case of creeks, a 10 m buffer zone should be created. No activities, such as harvesting, should take place within these zones. In addition to these measures the identification and promotion of alternative resources for building is required.

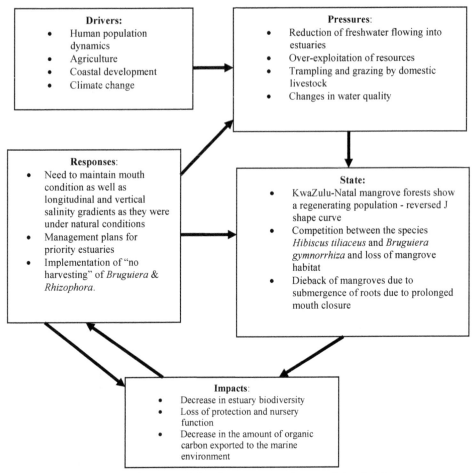

Fig. 9. Summary of DPSIR framework for the mangrove forests of South Africa.

5. Conclusion

Matrix modelling has allowed us to determine how much of a mangrove forest can be harvested while still maintaining a viable population. These data must be included in any management plan which includes the continual use of the forests as a wood resource for the local communities. The model presented here can be used by managers at other forests but growth data would need to be collected first as vital rates presented here will differ to other mangrove forests.

6. Acknowledgements

The authors would like to thank the National Research Foundation and the Nelson Mandela Metropolitan University for the funding of this study. Prof. Guy Bate and Dr. Taryn Riddin for reviewing and adding value to the paper.

7. References

Ajonina, G.N. (2008) *Inventory and Modeling Mangrove Forest Stand Dynamics Following Different Levels Of Wood Exploitation Pressures In The Douala-Edea Atlantic Coast Of Cameroon, Central Africa.* PhD Thesis-Albert-Ludwigs-Universität, Freiburg im Breisgau, Germany. 215 pg

Alongi, D.M. (2008) Mangrove forests: Resilience, protection from tsunamis, and responses to global climate change. *Estuarine Coastal and Shelf Science*, Vol. 76, No. 1 (January 2008), pp 1-13

Alongi, D.M. (2009) *Energetics of Mangroves.* Springer Science + Business Media B.V. ISBN-13: 978-1402042706, New York, United States of America

Ashton, E.C. & Macintosh, D.J. (2002) Preliminary assessment of the plant diversity and community ecology of the Sematan mangrove forest, Sarawak, Malaysia. *Forest Ecology and Management*, Vol.166, No. 1-3 (August 2002), pp 111-129

Atkins, J.P., Burdon, D., Elliott, M. & Gregory, A.J. (2011) Management of the marine environment: Integrating ecosystem services and societal benefits with the DPSIR framework in a systems approach. *Marine Pollution Bulletin*, Vol. 62, No. 2 (February 2011), pp 215-226

Bandaranyake, W.M. (1998) Traditional and medicinal uses of mangroves. *Mangroves and Salt Marshes*, Vol.2, No. 3 (February 1998), pp 133–148

Begg, G.W. (1984) The Estuaries Of Natal: Part 2. The Natal Regional and Planning Commission: Durban.

Bidone, E.D. & Lacerda, L.D. (2004) The use of the DPSIR framework to evaluate sustainability in coastal areas. Case study: Guanabara Bay basin, Rio de Janeiro, Brazil. *Regional Environmental Change*, Vol. 4, No. 1 (March 2004), pp 5-16

Bosire, J.O., Kairo, J.G., Kazungu, J., Koedam, N. & Dahdouh-Guebas, F. (2008) Spatial and temporal regeneration dynamics in *Ceriops tagal* (Perr.) C.B. Rob. (Rhizophoraceae) mangrove forests in Kenya. *Western Indian Ocean Journal of Marine Science*, Vol. 7, No. 1, pp 69-80

Boyce, M.S., Haridas, C.V., Lee, C.T. & NCEAS Stochastic Demography Working Group. (2006) Demography in an increasingly variable world. *Trends in Ecology and Evolution*, Vol. 21, No. 3 (March 2006), pp 141-148

Breen, C.M. & Hill, B.J. (1969) A mass mortality of mangroves in the Kosi Estuary. *Transactions of the Royal Society of Southern Africa*, Vol. 38, pp 285-303

Bruton, M.N. (1980) An outline of the ecology of the Mgobezeleni Lake System at Sodwana, with emphasis on the mangrove community. In: *Studies on the Ecology of Maputaland*, Bruton, M.N., Copper, K.H. (eds) pp (408-426), Cape and Transvaal Printers, Cape Town

Caswell, H. (2001) *Matrix Population Models: Construction, Analysis And Interpretation.* Second edition. Sinauer Associates, Massachusetts, United States of America

Caswell, H. (2009) Stage, age and individual stochasticity in demography. *Oikos*, Vol. 118, No. 12 (December 2009), pp 1763-1782

Choudhury, J.K. (1997) *Sustainable management of coastal mangrove forest development and social needs.* Proceedings of XI World Forestry Congress, Antalya – Turkey, October 1997

Clarke P.J. (1995), The population dynamics of the mangrove *Avicennia marina*; demographic synthesis and predictive modelling, *Hydrobiologia*, Vol. 295, No. 1-3 (January 1995), pp. 83–88

Clarke, P.J. & Kerrigan, R.A. & Westphal, C.J. (2001) Dispersal potential and early growth in 14 tropical mangroves: do early life history traits correlate with patterns of adult distribution? *Journal of Ecology*, Vol. 89, No. 4 (August 2001), pp 648-659

Cole, T.G., Ewel, K.C. & Devoe, N.N. (1999) Structure of mangrove trees and forests in Micronesia. *Forest Ecology and Management*, Vol. 117, No. 1-3 (May 1999), pp 95-109

Dahdouh-Guebas, F., Verneirt, M., Tack, J. F. & Koedam, N. (1997). Food preferences of *Neosarmatium meinerti* de Man (Decapoda: Sesarminae) and its possible effect on the regeneration of mangroves. *Hydrobiologia*, Vol. 347, No. 1-3 (March 1997), pp 83-89

Dahdouh-Guebas, F., Giuggioli, M., Oluoch, A., Vannini, M. & Cannicci, S. (1999). Feeding habits of non-ocypodid crabs from two mangrove forests in Kenya. *Bulletin of Marine Science*, Vol. 64, No. 2 (March 1999), 291-297

Dahdouh-Guebas, F., Pottelbergh, I., Kairo, J.G., Cannicci, S. & Koedam, N. (2004) Human-impacted mangroves in Gazi (Kenya): predicting future vegetation based on retrospective remote sensing, social surveys, and tree distribution. *Marine Ecology Progress Series*, Vol. 272 (May 2004), pp 77-92

Dahdouh-Guebas, F., Hettiarachchi, S., Lo Seen, D., Batelaan, O., Sooriyarachchi, S., Jayatissa, L.P. & Koedam, N. (2005) Transitions in ancient inland freshwater resource management in sri lanka affect biota and human populations in and around coastal lagoons. *Current Biology*, Vol. 15, No 6 (March 2005), pp 579–586

Day, J.H. (1980) What is an estuary? *South African Journal of Science*, Vol. 76, pp 198.

Emmerson, W., Cannicci, S. & Porri, F. (2003) New records for *Parasesarma leptosoma* (Hilgendorf, 1869) (Crustacea: Decapoda: Brachyura: Sesarmidae) from mangroves in Mozambique and South Africa. *African Journal of Zoology*, Vol. 38, No. 2 (January 2003), pp 351-355

Emmerson, W.D. & Ndenze, T.T. (2007) Mangrove tree specificity and conservation implications of the arboreal crab *Parasesarma leptosoma* at Mngazana, a mangrove estuary in the Eastern Cape, South Africa. *Wetlands Ecology and Management*, Vol. 15, No. (February 2007), pp 13–25

Ewel, K.C., Zheng, A., Pinzon, Z.S. & Bourgeols, J.A. (1998) Environmental effects of canopy gap formation in high-rainfall mangrove forests. *Biotropica*, Vol. 30, No. 4 (December 1998), pp 510-518

FAO (1994) Mangrove Forests Management Guidelines, FAO forestry paper 117, Rome, pp.169-191.

Fondo, E.N. & Martens, E.E. (1998) Effects of mangrove deforestation on macrofaunal densities, Gazi Bay, Kenya. *Mangroves and Salt Marshes*, Vol. 2, No. 2 (June 1998), pp 75-83

Gilbert, A.J. & Janssen, R. (1998) Use of environmental functions to communicate the values of a mangrove ecosystem under different management regimes. *Ecological Economics*, Vol. 25, No. 3 (June 1998), pp 323-346

Giri, C., Ochieng, E., Tieszen, L.L., Zhu, Z., Singh, A., Loveland, T., Masek, J. & Duke, N. (2011) Status and distribution of mangrove forests of the world using earth observation satellite data. *Global Ecology and Biogeography*, Vol. 20, No. 1 (December 2010), pp 154-159

Harun-or-Rashid, S., Biswas, S.R., Bocker, R. & Kruse, M. (2009) Mangrove community recovery potential after catastrophic disturbances in Bangladesh. *Forest Ecology and Management*, Vol. 257, No. 3 (February 2009), pp 923–930

Hogarth, P.J. (1999) *The Biology of Mangroves*. Oxford University Press, ISBN-0198502222, New York, United States of America

Kairo, G.K., Dahdou-Guebas, F., Gwada, P.O., Ochieng, C. & Koedam, N., (2002) Regeneration status of mangrove forests in Mida Creek, Kenya: a compromised or secured future? *Ambio*, Vol. 31, No. 7-8 (), pp 562-568

Kairo, J.G., Lang'at, J.K.S., Dahdouh-Guebas, F., Bosire, J. & Karachi, M. (2008) Structural
    development and productivity of replanted mangrove plantations in Kenya. *Forest
    Ecology and Management*, Vol. 255, No. 7 (April 2008), pp 2670-2677
Kathiresan, K. & Bingham, B.L. (2001) Biology of Mangroves and Mangrove Ecosystems.
    Advances in Marine Biology, Vol. 40, pp 81-251
Krauss, K.W., Lovelock, C.E., McKee, K.L., López-Hoffman, L., Ewe, S.M.L. & Sousa, W.P.
    (2008) Environmental drivers in mangrove establishment and early development:
    A review. *Aquatic Botany*, Vol. 89, No. 2 (August 2008), pp 105–127
Laegdsgaard, P. & Johnson, C. (2001) Why do juvenile fish utilise mangrove habitats? *Journal of
    Experimental Marine Biology and Ecology*, Vol. 257, No. 2 (March 2001), pp 229-253
López-Hoffman, L., Monroe, I.E., Narváez, E., Martínez-Ramos, M. & Ackerly, D.D. (2006)
    Sustainability of mangrove harvesting: how do harvesters' perceptions differ from
    ecological analysis? *Ecology and Society*, Vol. 11, No. 2 (July 2006): 14
Machiwa, J. F. & Hallberg, R. O. (2002). An empirical model of the fate of organic carbon in a
    mangrove forest partly affected by anthorpogenic activity. *Ecological Modelling*, Vol.
    147, No. 1 (January 2002), pp 69-83
Macintosh, D.J. & Ashton, E.C. (2004). Principles for a Code of Conduct for the
    Management and Sustainable use of Mangrove Ecosystems. Prepared for World
    Bank, ISME, cenTER Aarhus.
Maxim, L., Spangenberg, J.H. & O'Connor, M. (2009) An analysis of risks for biodiversity under
    the DPSIR framework. *Ecological Economics*, Vol. 69, No. 1 (November 2009), pp 12-23
Moll, E.J., Ward, C.J., Steinke, T.D. & Cooper, K.H. (1971) Our mangroves threatened.
    African Wildlife, Vol. 25, No. 3 (1971), pp 103-107
Mumby, P.J., Edwards, A.J., Lez, J.E.A., Lindeman, K.C., Blackwell, P.G., Gall, A.,
    Gorczynska, M.I., Harborne, A.R., Pescod, C.L., Renken, H., Wabnitz, C.C.C. &
    Llewellyn, G. (2003) Mangroves enhance the biomass of coral reef fish communities
    in the Caribbean. *Nature*, Vol. 427, No (2003), pp 533-536
Nagelkerken, I., Blaber, S.J.M., Bouillon, S., Green, P., Haywood, M., Kirton, L.G., Meynecke,
    J.O., Pawlik, J., Penrose, H.M., Sasekumar, A. & Somerfield, P.J. (2008) The habitat
    function of mangroves for terrestrial and marine fauna: A review. *Aquatic Botany*,
    Vol. 89, No. 2 (August 2008), pp 155-185
Omann, I., Stocker, A. & Jgeär, J. (2009) Climate change as a threat to biodiversity: An
    appliclation of the DPSIR approach. *Ecological Economics*, Vol. 69, No. 1 (November
    2009), pp 24-31
Owen-Smith, N. (2007) *Introduction to Modeling in the Wildlife and Resource Conservation*.
    Wiley-Blackwell Publishing. Oxford, United Kingdom.
Porte, A. & Bartelink, H.H. (2002) Modelling mixed forest growth: a review of models for
    forest management. *Ecological Modeling*, Vol. 150, No. 1-2 (April 2002), pp 141-188.
Rabinowitz, D. (1978) Early growth of mangrove seedlings in Panama, and a hypothesis
    concerning the relationship of dispersal and zonation. *Journal of Biogeography*, Vol.
    5, No. 2 (June 1978), pp 113-133.
Raimondo, D.C. & Donaldson, J.S. (2003) Responses of cycads with different life histories to
    the impact of plant collecting: simulation models to determine important life
    history stages and population recovery times. *Biological Conservation*, Vol. 111, No. 3
    (June 2003), pp 345-358.
Rajkaran, A. & Adams, J.B. (2007) Mangrove litter production and organic carbon pools in
    the Mngazana Estuary, South Africa. *African Journal of Aquatic Sciences*. Vol. 32, No.
    1 (January 2007): 17–25

Rajkaran, A., Adams, J.B. & Taylor, R. (2009) Historic and recent state (2006) of mangroves in small estuaries from Mlalazi to Mtamvuna in Kwazulu-Natal, South Africa. *Southern Forests*. Vol. 71, No. 4 (April 2009), pp 287-296

Rajkaran,   A. & Adams, J.B. (2010) The implications of harvesting on the population structure and sediment characteristics of the mangroves at Mngazana Estuary, Eastern Cape, South Africa. *Wetlands Ecology and Management*, Vol. 18, No. 1 (July 2010), pp 79-89.

Rajkaran, A. (2011) *A status assessment of mangrove forests in South Africa and the utilization of mangroves at Mngazana Estuary*. Ph.D. Thesis, Nelson Mandela Metropolitan University. 140 pg

Rockwood, L.L. (2006) *Introduction to Population Ecology*. Blackwell Publishing. Oxford, United Kingdom.

Saifullah, S.M., Shaukat, S. & Shams, S. (1994) Population structure and dispersion pattern in mangroves of Karachi, Pakistan. *Aquatic Botany*, Vol. 47, No. 3-4 (March 1994), pp 329-340

Sherman, R.E., Fahey, T.J. & Battles, J.J. (2000) Small-scale disturbance and regeneration dynamics in a neotropical mangrove forests. *Journal of Ecology*, Vol. 88, No. 1 (February 2000), pp 165-178

Silvertown, J.W. & Charlesworth, D. (2001) *Introduction to Plant Population Biology*. Blackwell Science Oxford, United Kingdom.

Skov, M. W. & Hartnol, R. G. (2002). Paradoxical selective feeding on a low-nutrient diet: why do mangrove crabs eat leaves? *Oecologia*, Vol. 131, No. 1 (March 2002), pp 1-7

Smith, T. J., Boto, K. G., Frusher, S. D. & Giddins, R. L. (1991). Keystone species and mangrove forest dynamics: the influence of burrowing by crabs on soil nutrient status and forest productivity. *Estuarine, Coastal and Shelf Science*, Vol. 33, No. 5 (November 19991), pp 419-432

Spalding, M., Kainuma, M. & Collins, L. (2010) *World Atlas of Mangroves*. The International Society for Mangrove Ecosystems, ISBN-13: 978-1844076574, Okinawa, Japan.

Thampanya, U., Vermaat, J.E., Sinsakul, S. & Panapitukkul N (2006) Coastal erosion and mangrove progradation of Southern Thailand. *Estuarine, Coastal and Shelf Science*, Vol. 68, No. 1-2 (June 2006), pp 75-85

Traynor, C.H. & Hill, T. (2008) Mangrove utilisation and implications for participatory forest management, South African. *Conservation and Society*, Vol. 6, No. 2 (March 2008): 109-116

Walters, B.B. (2005) Ecological effects of small-scale cutting of Philippine mangrove forests. *Forest Ecology and Management* 206, No. 1-3 (February 2005), pp 331–348

Walters, B.B., Rönnbäck, P., Kovacs, J.M., Crona, B., Hussain, S.A., Badola, R., Primavera, J.H., Barbier, E. & Dahdouh-Guebas, F. (2008) Ethnobotany, socio-economics and management of mangrove forests: A review. *Aquatic Botany*, Vol. 89, No. 2 (August 2008), pp 220-236

Wilkie, M.L. & Fortuna, S. (2003) Status and Trends in Mangrove Area Extent Worldwide. By. Forest Resources Assessment Working Paper No. 63. Forest Resources Division. FAO, Rome. (Unpublished).

Whitfield, A.K. (1992) A characterisation of southern African estuarine systems. *Southern African Journal of Aquatic Sciences*, Vol. 12, No. 1-2 (January 1993), pp 89–103

10

A Decision-Support Model for Regulating Black Spruce Site Occupancy Through Density Management

P. F. Newton
Canadian Wood Fibre Centre, Canadian Forest Service, Natural Resources Canada
Canada

1. Introduction

Regulating site occupancy through stand density management has been a cornerstone of silvicultural practice since it was first introduced in forestry by Reventlow in 1879 (Pretzsch, 2009). Density management continues to be a dominant intensive forest management practice throughout boreal and temperate forest regions (e.g., Canada (CCFM, 2009) and Finland (Peltola, 2009) treat over 500,000 ha annually). Operationally, density management consists of manipulating initial planting densities at the time of establishment (initial espacement; IE) and (or) reducing stand densities during subsequent stages of stand development (e.g., precommercial thinning (PCT) at the sapling stage, and (or) commercial thinning (CT) at the semi-mature stage). As documented by numerous case studies, density management can result in a wide array of benefits at the tree, stand and forest levels. These include increased growth and resultant yields leading to enhanced end-products (e.g., Kang et al., 2004), attainment of early stand operability status (e.g., Erdle, 2000), reduced density-dependent mortality losses (e.g., Pelletier & Pitt, 2008), increased spatial and structural uniformity resulting in lower extraction, processing and manufacturing costs (e.g., Tong et al., 2005), and increased carbon sequestration rates (e.g., Nilsen & Strand, 2008). Density management also has consequential effects on other important non-timber values. These include regulating the production of coarse woody debris to met wildlife habitat requirements (e.g., pine marten (*Martes americana*) (Sturtevant et al., 1996)), provision of thermal protection and hiding requirements for ungulates by regulating stand structure (e.g., elk (*Cervus elaphus nelsonii*) and mule deer (*Odocoileus hemionus*) (Smith and Long, 1987)), controlling successional pathways in order to prevent the establishment and development of ericaceous shrub species (e.g., Lindh and Muir, 2004), and increasing biodiversity (e.g., Verschuyl et al., 2011). Although thinning effects are largely positive in nature, inappropriate treatments can have serious detrimental implications. These include (1) PCT treatments which result in an extended period of openness in which individual trees are allowed to build up extensive crowns resulting in an increase in juvenile wood production and larger knot sizes (e.g., Tong et al., 2009), and (2) CT treatments which are implemented within structurally unstable stands resulting in increased mortality during high wind or heavy ice and snow events.

Determination of the optimal density management regime for a given objective is a complex process given the multitude of variables that a forest manager needs to consider. For example, deciding on initial establishment densities, the timing of thinning entries and associated removables, discount and interest rates, and fixed and variable cost values. Furthermore, the selected regime must be considered within the broader regulatory framework which can impose additional constraints on the decision-making process (e.g., specific minimum pre-treatment tree size and basal area requirements before CT treatments can be implemented (McKinnon et al., 2006)). Fortunately, however, the complexity of decision-making has been greatly reduced for traditional volumetric-based objectives with the advent of stand density management diagrams (SDMDs; Ando, 1962; Drew & Flewelling, 1979; Jack & Long, 1996; Newton, 1997).

Briefly, SDMDs are graphical decision-support tools that are used to determine the density management regime required for the realization of a specified mean tree size or volumetric yield objective. Recently, in order to address the evolving paradigm shift in management focus from a singular volumetric yield maximization objective to a focus on a multitude of diverse objectives, including the end-product quality (Barbour & Kellogg, 1990), product value maximization (Emmett, 2006), bioenergy and carbon sequestration potential, and ecosystem services, the SDMD modeling framework was expanded. Specifically, Newton (2009) introduced the modular-based structural stand density management model (SSDMM) for jack pine (*Pinus banksiana* Lamb.) stand-types. The model has a hierarchical design in which 6 integrated estimation modules collectively enable the estimation of volumetric productivity, log distributions, product volumes and values, and fibre attributes, for a given density management regime, site quality, and cost profile.

The objectives of this study were to describe the upland black spruce (*Picea mariana* (Mill.) BSP) variant of the modular-based SSDMM and demonstrate its utility in designing density management regimes within an operational context. More specifically, the stand-level examples are placed within the broader context of sustainable management at the landscape level in which a portion of the productive forest land base is allocated and managed for timber related objectives (i.e., early operability within natural-origin stands, and production of enhanced end-products within plantations) and the remainder, for non-timber related objectives (i.e., production of coarse woody debris (CWD) for maintenance of wildlife habitat).

2. Methods

2.1 Modular-based SSDMM for upland black spruce stands

The SSDMM for upland black spruce stands was developed by expanding the dynamic SDMD modelling framework through the incorporation of diameter, height, log-type, biomass, carbon, product and value distribution, and wood quality recovery modules (Figure 1). Analytically, the principal steps involved the development of a dynamic SDMD and the subsequent incorporating of (1) a parameter prediction equation (PPE) system for diameter distribution recovery, (2) a composite height-diameter prediction equation for height estimation, (3) a composite taper equation for recovering log product distributions and calculating stem volumes, (4) composite biomass equations for estimating above ground components and their carbon mass equivalents, (5) sawmill-specific product recovery and associated product value functions, and (6) composite wood density and maximum mean branch diameter equations. Computationally, Module A (Dynamic SDMD) provides a set of annual stand-level variables which are required as input to Modules B-F. Module B utilizes the PPE system and the composite height-diameter function to recover the grouped-diameter frequency distribution and estimate corresponding tree heights for each diameter class (Diameter and Height Recovery Module), and similar to Module A, provides prerequisite input to the remaining modules. The taper equation is used to derive estimates of the upper stem diameters for each tree within each diameter class from which the number of sawlogs and pulplogs, residual tip volumes, and merchantable and total stem volumes, are calculated (Taper Analysis and Log Estimation Module). The composite biomass equations are used to predict masses and carbon equivalents for each above-ground component (Biomass and Carbon Estimation Module). The product recovery and value functions are used to predict sawmill-specific (stud mill (SM) and randomized length mill (RLM)) chip and lumber volumes and associated market-based monetary values (Product and Value Estimation Module). The composite fibre attribute functions are used to estimate mean wood density for merchantable-sized ($\geq$ 10 cm diameter classes) trees, and the mean maximum branch diameter within the first 5 m sawlog for trees $\geq$ 15.1 cm in diameter (Fibre Attribute Estimation Module). Refer to Newton (2012a) for a complete description of the approach used in the development and calibration of the modular-based SSDMM for upland black spruce stands.

Given the model's complexity and the computation burden associated with its use, an algorithmic analogue was developed in the Visual Basic (VB.NET (Ver. 1.1); Microsoft Corporation) programming language. Denoted, Croplanner, the program predicts and tabulates site-dependent annual and rotational diameter-class and stand-level estimates of volumetric yields, log distributions, biomass and carbon outcomes, recoverable products and associated values by sawmill-type, economic efficiency profiles and fibre attributes, for 3 density management regimes per simulation. The user is required to specify the following information for each simulation: (1) provincial region (e.g., Ontario); (2) stand-type (natural origin or plantation); (3) simulation year; (4) site quality (site index); (5) rotational age; (6) establishment densities; (7) expected ingress during the establishment period (n., applicable to plantations only); (8) merchantable specifications (i.e., length and upper threshold diameters for pulp and saw logs, and merchantable top diameter); (9) interest and discount rates; (10) operability targets (i.e., number of merchantable trees per cubic metric of merchantable wood, and total merchantable volume per unit area); (11) establishment costs (e.g., fixed site assessment or preparation expenses and planting costs); (12) genetic worth effects and selection ages (n., applicable to plantations only); (13) operational adjustment factors; (14) product degrade estimates; (15) variable cost estimates accounting for stumpage and renewal charges, harvesting, transportation and manufacturing expenses at the time of harvest; and (16) regime-specific thinning treatments and associated costs (i.e., time of entry (stand age), type of thinning (PCT or CT), removal densities (stems/ha) or basal area (%) reductions, and fixed and variable thinning cost values).

For each year, the program recovers the grouped-diameter frequency distribution and for each recovered diameter class, calculates height, number of pulp and saw logs, merchantable and total volumes, biomass and carbon equivalents for each above-ground component (bark, stem, branch and foliage), sawmill-specific recoverable chip and lumber volumes and associated monetary values, and mean tree fibre attributes. Cumulative stand-level values and performance indices are subsequently derived. The output is presented in both tabular and graphical formats and consists principally of a traditional SDMD graphic, regime- specific annual estimates at the individual diameter-class level and stand-levels, regime-specific treatment and rotational summaries, and across-regime rotational comparisons. The comparisons employ a comprehensive set of performance indices which include measures of (1) overall productivity as measured by the mean annual merchantable

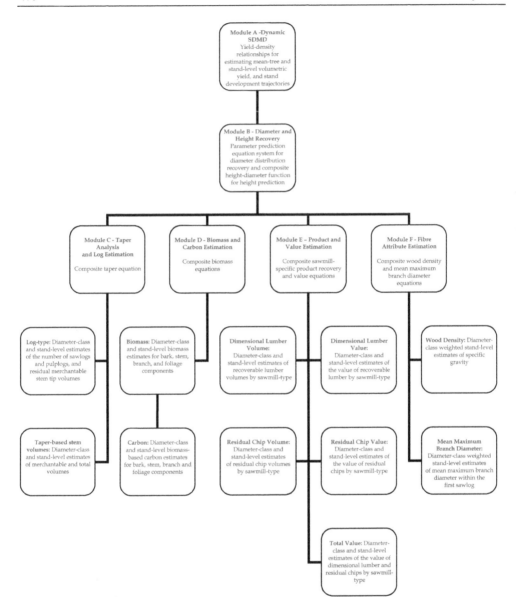

Fig. 1. Schematic illustration of the modular-based SSDMM.

volume increment (m³/ha/yr), mean annual biomass increment (t/ha/yr) and mean annual carbon increment (t/ha/yr), (2) log production in terms of the percentage by sawlogs produced, (3) end-products recovered as quantified by the percentage of lumber volume produced by each sawmill type, (4) economic efficiency based on land expectation values (i.e., the maximum an investor could pay for bare land to achieve a specified rate of return (discount rate)) of a given manipulated regime relative to the control regime for each sawmill type), (5) optimal site occupancy (number of years that a size-density trajectory was within an optimal production zone as delineated by relative density indices of 0.32 and 0.45 (Newton, 2006)), (6) stand stability as reflected by the mean height/diameter ratio for trees within the dominant crown class, (7) fibre quality attributes as summarized by mean wood density and mean maximum branch diameter, (8) accelerated operability based on the reduction in the number of years that a stand took to reach harvestable status as defined by target piece size and merchantable yield thresholds, and (9) time to full occupancy as quantified by the number of years required to reach initial crown closure status.

2.2 Simulations
The treatment regimes as stated within an operational forest management plan are used to exemplify the utility of model. Specifically, the silvicultural matrix presented in the 2009-2019 forest management plan developed for the Romeo Malette Forest in the Timmins District of the Northeastern Region of Ontario, Canada, by Tembec Inc. (Anonymous, 2009), was used. These ecosite-specific treatment regimes reflect best management practices for a given stand and forest management objective as defined within the NEBIE silvicultural intensity framework (Bell et al., 2008).

For the natural regenerated stand-type (forest unit SP1 (Ecosite 2)), an extensive silvicultural intensity employing an early operability objective, was evaluated. For the plantations (forest unit SP1 (Ecosite 5f)), an elite silvicultural intensity with an enhanced end-product value objective, was evaluated. These objectives reflect ongoing discussions regarding the management of boreal conifers in the central portion of the Canadian Boreal Forest Region: (1) implementing PCT treatments within density-stressed natural-origin stands in order to shorten the time to operability status; and (2) employing CT treatments within genetically-improved plantations so that merchantable volume losses normally attributed to density-dependent mortality at the later stages of stand development are minimized, and reducing the technical rotation age in regards to the production of high quality wood products.

The protocol for implementing the CT treatments followed the provincial recommendations as espoused by McKinnon et al. (2006). Specifically, preferable CT density management regimes are those which (1) increased mean tree size without incurring declines in stand volume growth, (2) do not unacceptably increase the risk of volume losses to wind, snow, insects, and disease, and (3) minimize the rate of density-dependent mortality within the merchantable-sized classes during the later stages of stand development thus enabling the recovery of some of the expected merchantable volume losses through thinning. Operationally, the CT treatment should occur within previously density regulated stands which are approximately 15-20 yrs from rotation age. The CT treatment should reduce basal areas by a maximum of 30-35% from an initial minimum basal area of 25 m^2/ha and be implemented only when density-dependent mortality is occurring or imminent within the merchantable-sized classes. Lastly, CT treatments should only occur within stands where the mean live crown ratio exceed 35%. Table 1 provides a summary of the input parameters required to run these scenarios with the Croplanner algorithm.

3. Results and discussion

3.1 Extensive silviculture: Natural-origin black spruce stand-types subjected to PCT
The resultant mean volume-density trajectories for the natural-origin black spruce stands within the context of the traditional SDMD graphical format are illustrated in Figure 2. It is

Input Parameter (unit)	Stand-type and Treatment					
	Natural-origin Stands subjected to PCT			Plantations subjected to IE+PCT+CT with genetic worth effects		
	Regime 1 – Control	Regime 2 – PCT	Regime 3 – PCT	Regime 1 – Control	Regime 2 – PCT	Regime 3 – PCT+CT
Silvicultural intensity	Extensive			Elite		
Objective	Early operability			End-product value		
Simulation year	2011	2011	2011	2011	2011	2011
Site index (Carmean et al., 2006)	16	16	16	18	18	18
Rotation age (yr)	80	80	80	50	50	50
Initial density (stems/ha)	5000	5000	5000	2750	2750	2750
Ingress density (stems/ha)	-	-	-	0	0	0
Merchantable specifications						
Pulplog length (m)	2.59	2.59	2.59	2.59	2.59	2.59
Pulplog minimum diameter (cm)	10	10	10	10	10	10
Sawlog length (m)	5.03	5.03	5.03	5.03	5.03	5.03
Sawlog minimum diameter (cm)	14	14	14	14	14	14
Merchantable top diameter (cm)	4	4	4	4	4	4
Rates						
Interest rate (%)	2	2	2	2	2	2
Discount rate (%)	4	4	4	4	4	4
Operability Targets						
Piece-size (stems/m³)	10	10	10	10	10	10
Merchantable yield (m³/ha)	130	130	130	200	200	200
Site preparation ($/ha)	100	100	100	300	300	300
Planting ($/seedling)	-	-	-	0.6	0.6	0.6
Genetic worth (%)	-	-	-	15	15	15
Selection age (yr)	-	-	-	15	15	15
Operational adjustment factor (%)	1	1	1	1	1	1
Product degrade (%)	15	5	5	15	10	5
Variable costs for harvesting, stumpage, renewal, transportation and manufacturing ($/m³)	100	80	80	75	65	55
PCT Treatments						
Time of treatment (yr)	-	14	14	-	13	13
Number of trees removed (stems/ha)	-	1943	2943	-	907	907
Fixed cost of PCT ($/ha)	-	300	300	-	300	300
CT Treatment						
Time of treatment (stems/ha)	-	-	-	-	-	30
Number of trees removal (stems/ha)	-	-	-	-	-	604
Fixed cost of CT ($/ha)	-	-	-	-	-	100
Variable cost for harvesting, stumpage, transportation and manufacturing for volume removed ($/m³)	-	-	-	-	-	65

Table 1. Stand-type specific input parameters used in the Croplanner simulations.

instructive to familiarize oneself with the overall structure of the diagram, particularly, in relation to the static and dynamic components. Essentially, the yield-density isolines are used for positioning a given stand in the size-density space and deriving corresponding yield estimates. The size-density trajectories in combination with the isolines provide a graphical pictorial of overall stand dynamics (density changes due to thinning treatments and density-dependent and independent mortality) in addition to enabling users to derived structural characteristics at various key phases of stand development, through interpolation. For example, the intersection of the size-density trajectories with the diagonal line denoting crown closure status indicated that the stand thinned to a residual density of 3000 trees (stems/ha; Regime 2) re-attained crown closure status by an age of 18 yr whereas the stand thinned to a residual density of 2000 trees (stems/ha; Regime 3) re-attained crown closure status by an age of 22 yr. Knowing the period of time a stand is open-grown is an important metric when attempting to control early branch development within the lower portion of the stem through density regulation.

The graphic also shows that at an approximate mean dominant height value of 10 m, the stands enter a period of accelerated self-thinning, as evident from the degree of curvature of the size-density trajectories. The degree of self-thinning was most pronounced in the control stand and less so for the PCT treated stands. Numerically, from the time of treatment to rotation, the unthinned control stand lost 3373 trees (stems/ha; Regime 1) compared with only 1746 trees (stems/ha) for Regime 2, and 1124 trees (stems/ha) for Regime 3. By the time the stands reached rotation age (80 yr) they were positioned just below the 20 m mean dominant height isoline. The control stand was just below the 18 cm quadratic mean diameter isoline, just above the 0.9 relative density index isoline, and just below the 35% mean live crown ratio isoline. For the thinned stand PCT to a residual density of 3000 stems/ha (Regime 2), the trajectory terminated at a position that was slightly above the 18 cm quadratic mean diameter isoline, just above the 0.8 relative density index isoline, and slightly below the 35% mean live crown ratio isoline. Similarly, for the thinned stand PCT to a residual density of 2000 stems/ha (Regime 3), the trajectory terminated at a position that intersected the 20 cm quadratic diameter isoline, just above the 0.7 relative density index isoline, and intersected the 35% mean live crown ratio isoline. Although the graphic is very useful in terms of understanding and visualizing stand development, the algorithmic revision readily facilitates the estimation of a much broader array of yield, end-product, economic, and wood fibre attribute metrics (Table 2), and associated performance measures (Table 3), at various temporal scales (annual, periodic and rotational).

The thinning treatments resulted in an increase in the duration of the pre-crown-closure period by 4 and 8 yr for Regimes 2 and 3, respectively. Given that the dominant height of the stands would be in the 5.5 to 6.5 m range at time of re-closure, most of the branches within the first 5 m long sawlog would have been formed by then. As inferred by the minimal differential in mean maximum branch diameters at rotation between the stands (c.f., 2.65 cm versus a mean of 2.70 cm for the control and thinned stands, respectively; Table 3), suggest that this extended period of openness did not consequentially affect branch development within this economically-important portion of the stem. Comparing Regimes 2 and 3 against Regime 1, indicated that on the positive side, the PCT treatment (1) shorten the time to stand operability status by an average of 8 years, (2) produced trees of large mean size at rotation (i.e., average increases in mean volume of 32%), (3) increased the percentage of sawlogs produced by an average of 12%, and (4) enhanced overall structural stability (e.g., reducing the height/diameter ratio by an average of 10%). On the negative side, however, the single PCT treatment resulted in lower per unit yields for merchantable volume (average of 12% less), and biomass and carbon production (average of 18% less). Economically, however, the PCT treatments did result in gains in economic efficiency (an average of 88% increase) at the specified rotation age of 80 yr, irrespectively of sawmill type. These economic differences can be largely attributed to the lower product degrade values specified for the thinned stands, and to the assumed reduction in variable costs at the time of harvest arising from decreased harvesting and manufacturing expenses due to increased piece-size,

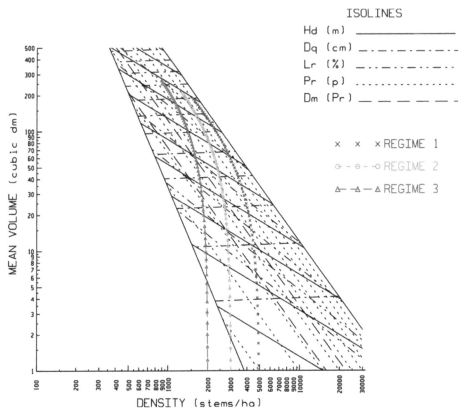

Fig. 2. Dynamic SDMDs for natural-origin upland black spruce stand-types managed under an extensive silvicultural intensity. Graphically illustrating (1) isolines for mean dominant height (Hd; 6-22 m by 2 m intervals), quadratic mean diameter (Dq; 4-26 cm by 2 cm intervals), mean live crown ratio (Lr; 35, 40, 50,..., 80%), and relative density index (Pr; 0.1-1.0 by 0.1 intervals), (2) the self-thinning line at a Pr = 1.0, and initial crown closure line (lower solid diagonal line); (3) lower and upper Pr values delineating the optimal density management window (Dm; 0.32 ≤ Pr ≤ 0.45); and (4) expected 80-yr size-density trajectories with 1 year intervals denoted for 3 user-specified density management regimes for stands situated on a medium site quality (site index = 16).

reduced size variation, and more uniform spatial patterns. In terms of provision of wildlife trees, the number of large standing snags (trees/ha), as approximated by the number of merchantable-sized abiotic trees which died during the last decade before harvest, was 41% less in the PCT stands as compared to the control stand.

Attribute (unit)	Regime 1 - Control	Regime 2 – PCT (thinning yields)	Regime 3 – PCT (thinning yields)
Mean dominant height (m)	19.9	19.9	19.9
Quadratic mean diameter (cm)	18	19	20
Basal area (m²/ha)	39	35	32
Mean volume per tree (dm³)	186	216	252
Total volume (m³/ha)	291	272 (4)	242(6)
Total merchantable volume (m³/ha)	276	257 (0)	229(0)
Density (stems/ha)	1561	1254 (1943)	876 (2943)
Relative density index (%/100)	0.91	0.81	0.66
Mean live crown ratio (%)	33	34	35
Number of pulplogs (logs/ha)	3149	2345 (-)	1531 (-)
Number of sawlogs (logs/ha)	724	846 (-)	782 (-)
Residual log tip volume (m³/ha)	53	39 (-)	28 (-)
Bark biomass (t/ha)	19	17 (-)	15 (-)
Stem biomass (t/ha)	188	164 (-)	139 (-)
Branch biomass (t/ha)	6	6 (-)	6 (-)
Foliage biomass (t/ha)	11	11 (-)	12 (-)
Total biomass (t/ha)	224	198 (-)	172 (-)
Bark carbon (t/ha)	10	9 (-)	8 (-)
Stem carbon (t/ha)	94	82 (-)	70 (-)
Branch carbon (t/ha)	3	3 (-)	3 (-)
Foliage carbon (t/ha)	5	5 (-)	6 (-)
Total carbon (t/ha)	112	99 (-)	86 (-)
Chip volume – SM (m³/ha)	127	110 (-)	90 (-)
Lumber volume – SM (m³/ha)	149	133 (-)	122 (-)
Chip volume – RLM (m³/ha)	108	94 (-)	77 (-)
Lumber volume – RLM (m³/ha)	166	148 (-)	134 (-)
Chip value – SM ($K/ha)	6	6 (-)	5 (-)
Lumber value – SM ($K/ha)	27	27 (-)	26 (-)
Total product value – SM ($K/ha)	33	33 (-)	31 (-)
Chip value – RLM ($K /ha)	6	5 (-)	4 (-)
Lumber value – RLM ($K /ha)	36	37 (-)	35 (-)
Total product value – RLM ($K/ha)	42	42 (-)	39 (-)
Land expectation value – SM ($K/ha)	1.2	2.7	2.6
Land expectation value RLM - ($K/ha)	3.2	4.7	4.3

Table 2. Rotational yield estimates for upland black spruce natural-origin stands subjected to PCT. Values in parenthesis denote yields derived from the PCT treatment (n., a dash line indicates an incalculable value).

Index (unit)	Regime 1 - Control	Regime 2 - PCT	Regime 3 - PCT
Mean annual volume increment (m³/ha/yr)	3.4	3.2	2.9
Mean annual biomass increment (t/ha/yr)	2.8	2.5	2.1
Mean annual carbon increment (t/ha/yr)	1.4	1.2	1.1
Percentage of sawlogs produced (%)	19	27	34
Percentage of lumber volume recovered - SM (%)	54	55	57
Percentage of lumber volume recovered - RLM (%)	61	61	64
Relative land expectation value – SM (%)	-	138	49
Relative land expectation value – RLM (%)	-	129	36
Duration of optimal site occupancy (%)	9	11	16
Mean height/diameter ratio (m/m)	103	96	90
Mean wood density (g/cm³)	0.48	0.49	0.49
Mean maximum branch diameter (cm)	2.65	2.68	2.72
Time to operability status (yr)	64	58	55
Time to initial crown closure (yr)	14	14	14
Age of crown re-closure post-PCT (yr)	-	18	22
Number of large standing snags (trees/ha)	332	228	165

Table 3. Stand-level performance indices for density-manipulated upland black spruce natural-origin stands subjected to PCT.

In summary, this specific simulation indicated that PCT resulted in (1) earlier stand operability status, (2) larger but fewer trees at rotation, (3) an increased in the duration of optimal site occupancy, (4) enhanced structural stability, (5) a decline in overall merchantable volume productivity, and (6) production of fewer wildlife trees.

3.2 Elite silviculture: Genetically-improved upland black spruce plantations subjected to PCT and CT

Similar to the PCT treatments within the natural-origin stands, the resultant mean volume-density trajectories for elite treatments are graphically illustrated within the context of the SDMD graphic (Figure 3). Table 4 lists the rotational and thinning yield estimates whereas Table 5 lists the resultant stand-level performance indices. Although self-thinning occurred within all 3 regimes indicating full occupancy had been achieved, the rate of density-dependent mortality increased with increasing planting density. The PCT treatments extended the period of openness by approximately 3 yr, however the effect of branch development was minimal (c.f., 2.65 cm for the control stand versus 2.69 and 2.72 cm for the PCT and PCT+CT stands, respectively). The trajectories also revealed that the thinned stands spent a greater portion of the rotation in the optimal site occupancy zone: 20% and 44% for the PCT and PCT+CT regimes, respectively, versus 12% for the control stand. This suggest that the thinned stands, particularly the stand that received a dual treatment (Regime 3), the rate of carbon sequestration and biomass production was close to an optimal level for a considerable portion of the rotation. Essentially, stands below the zone are not fully utilizing the site and consequently site resources are going unused in terms of forest biomass production (e.g., resource supply exceeds demand). Stands above the zone are over-occupying the site resulting in intensive asymmetric resource competition among local neighbors and subsequent mortality through self-thinning.

Further examination of the SDMD revealed that the size-density trajectories intersected the crown closure isoline slightly above the 4 m mean dominant height isoline. This corresponds to an age of 13 yr this site quality and represents the target PCT age. The yield-density isolines indicated that the stands were slightly above the 4 cm quadratic mean diameter isoline and the 0.1 relative density isoline, at the time of the PCT treatment. The corresponding interpolated mean volume, density and basal area values were 2.9 dm³, 2707 stems/ha and 4.0 m²/ha, respectively. Similarly, the size-density trajectories at the time of the CT treatment were slightly above the 12 m mean dominant height isoline, 14 cm quadratic mean diameter isoline, 0.5 relative density isoline, and the 40% live crown ratio isoline. The corresponding interpolated mean volume, density and basal area values were 77.9 dm³, 1604 stems/ha and 25.4 m2/ha, respectively. Accordingly, the stands would be candidates for CT treatments based on the guidelines given by McKinnon et al. (2006): CT candidate stands must have been previously managed in terms of density control treatments (e.g., IE with PCT), have a pretreatment basal area of greater than 25 m²/ha, a mean live crown ratio greater than 35%, and where density-dependent mortality within the merchantable size classes is imminent. In case of the PCT stands, this last requirement was projected to occur at an age of 31 yr.

The mean dominant height at rotation age was 17.2 m for all 3 plantations. Respectively, for Regimes 1, 2 and 3, the rotational values for mean live crown ratio were 33, 34 and 38% and cumulative merchantable volume were 284, 240 and 211 m³/ha. The CT treatment consisting of removing 35% (8.8 m²/ha) of basal area at age 30 resulted in a mid- rotation harvest of approximately 39 m³/ha of merchantable volume. Density-dependent mortality rates within the merchantable size classes of the CT stand was considerably lower than that within both the control and PCT stand during the post-CT period (c.f., 204 stems/ha within the PCT+CT stand versus 710 and 389 stems/ha within the control and PCT stands, respectively, over the 20 yr period). Although, relative to the control and PCT stand, the CT treatment resulted in larger but fewer trees of slightly inferior quality at rotation, the dual treatment did extended period of optimal site occupancy and substantially increased the economic worth of the stand at rotation. Relative to the control stand, the number of large standing snags (trees/ha) at rotation was approximately 39% and 73% less in the PCT and PCT+CT treated stands, respectively.

In summary, relative to the unthinned plantation, the thinning treatments resulted in (1) lower overall productivity in terms of merchantable volume (16 and 26% less for the PCT and PCT+CT plantations, respectively), and biomass and carbon production (8 and 11% less for the PCT and PCT+CT plantations, respectively), (2) extended the time to operability status by 6 and 12 yr for the PCT and PCT+CT plantations, respectively, (3) larger (mean volume) but fewer trees at rotation, (4) increased economic efficiency (36 and 54% less for the PCT and PCT+CT plantations, respectively), and (5) increased durations of optimal site occupancy (8 and 36% more for the PCT and PCT+CT plantations, respectively). With respect to the single core objective of increasing the production of high-value end-products through thinning, the results were not fully supportive. For the 2 mill configurations assessed, the thinned plantations produced lower volumes of chip (13 and 18% less for the PCT and PCT+CT plantations, respectively) and dimensional lumber products (9 and 20% less for the PCT and PCT+CT plantations, respectively). The removal of the merchantable-sized trees during the CT contributed to the decline in sawlogs and associated dimensional lumber volumes at rotation. In terms of product values, the thinned stands produced generally lower monetary values due to the decreased end-product volumes. However the differences were not large and in some cases were nil (c.f., product valves for the RLM configuration for the thinned versus control plantation (Table 4)). The largest benefit from thinning was in terms of an increase in economic efficiency as inferred from the ratio of land expectation values between the control and the treated plantations (Table 5). The lower product degrade values employed and the assumed lower variable costs arising from a more uniform piece-size distribution, largely contributed to this positive economic result.

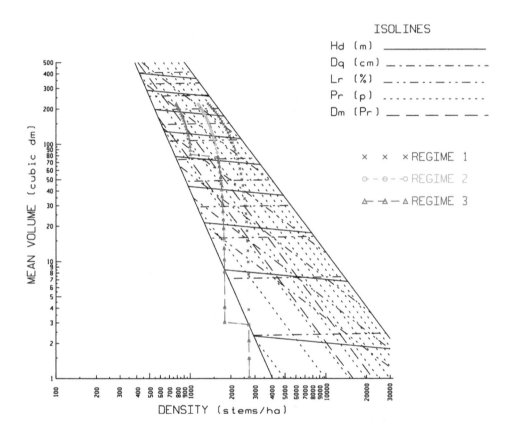

Fig. 3. Dynamic SDMD for genetically enhanced upland black spruce plantations managed under an elite silvicultural intensity. Graphically illustrating: (1) isolines for mean dominant height (Hd; 4-20 m by 2 m intervals), quadratic mean diameter (Dq; 4-26 cm by 2 cm intervals), mean live crown ratio (Lr; 35, 40, 50,…, 80%), relative density index (Pr; 0.1-1.0 by 0.1 intervals); (2) self-thinning line at a Pr = 1.0 and initial crown closure line (lower solid diagonal line); (3) lower and upper Pr values delineating the optimal density management window (Dm; $0.32 \leq Pr \leq 0.45$); and (4) expected 50 year size-density trajectories with 1 year intervals denoted for 3 user-specified density management regimes for plantations situated on a good site quality (site index = 18).

Attribute (unit)	Regime 1 - Control	Regime 2 - PCT (thinning yields)	Regime 3 – PCT+CT (thinning yields)
Mean dominant height (m)	17.2	17.2	17.2
Quadratic mean diameter (cm)	21	21	21
Basal area (m²/ha)	46	38 (1)	27 (1,9)
Mean volume per tree (dm³)	222	227	237
Total volume (m³/ha)	302	255 (2)	183 (2,43)
Total merchantable volume (m³/ha)	284	240 (0)	173 (0,39)
Density (stems/ha)	1358	1120 (907)	773 (907,604)
Relative density index (%/100)	0.89	0.74	0.53
Mean live crown ratio (%)	33	34	38
Number of pulplogs (logs/ha)	1982	1667 (0)	1203 (0,464)
Number of sawlogs (logs/ha)	909	824 (0)	643 (0,0)
Residual log tip volume (m³/ha)	42	35 (0)	25 (0,9)
Bark biomass (t/ha)	18	17 (0)	14 (0,3)
Stem biomass (t/ha)	164	144 (1)	110 (1,21)
Branch biomass (t/ha)	9	8 (1)	8 (1,3)
Foliage biomass (t/ha)	14	15 (2)	16 (2,5)
Total biomass (t/ha)	205	184 (5)	147 (5,31)
Bark carbon (t/ha)	9	8 (0)	7 (0,1)
Stem carbon (t/ha)	82	72 (1)	55 (1,10)
Branch carbon (t/ha)	4	4 (1)	4 (1,1)
Foliage carbon (t/ha)	7	8 (1)	8 (1,3)
Total carbon (t/ha)	103	92 (2)	74 (2,16)
Chip volume – SM (m³/ha)	123	107 (0)	79 (0,22)
Lumber volume – SM (m³/ha)	131	120 (0)	94 (0,11)
Chip volume – RLM (m³/ha)	106	92 (0)	68 (0,20)
Lumber volume – RLM (m³/ha)	146	133 (0)	105 (0,13)
Chip value – SM ($K/ha)	7	6 (0)	5 (0,1)
Lumber value – SM ($K/ha)	25	24 (0)	21 (0,2)
Total product value – SM ($K/ha)	32	30 (0)	26 (0,3)
Chip value– RLM ($K /ha)	4	4 (0)	3 (0,1)
Lumber value– RLM ($K /ha)	34	34 (0)	29 (0,5)
Total product value – RLM ($K/ha)	38	38 (0)	32 (0,5)
Land expectation value – SM ($K/ha)	3.7	5.5	6.4
Land expectation value - RLM - ($K/ha)	7.3	8.9	9.8

Table 4. Rotational yield estimates for upland black spruce plantations established at fixed IE levels subjected to PCT and CT treatments with genetic worth effects incorporated. Values in parenthesis denote yields derived from the thinning treatment(s) (ordered by time of treatment).

Index (unit)	Regime 1 - Control	Regime 2 - PCT	Regime 3 - PCT+CT
Mean annual volume increment (m³/ha/yr)	5.7	4.8	4.2
Mean annual biomass increment (t/ha/yr)	4.1	3.8	3.7
Mean annual carbon increment (t/ha/yr)	2.1	1.9	1.8
Percentage of sawlogs produced (%)	31	33	28
Percentage of lumber volume recovered - SM (%)	52	53	51
Percentage of lumber volume recovered - RLM (%)	58	59	57
Relative land expectation value – SM (%)	-	49	74
Relative land expectation value – RLM (%)	-	22	34
Duration of optimal site occupancy (%)	12	20	48
Mean height/diameter ratio (m/m)	72	72	70
Mean wood density (g/cm³)	0.48	0.49	0.50
Mean maximum branch diameter (cm)	2.65	2.69	2.72
Time to operability status (yr)	35	41	47
Time to initial crown closure (yr)	13	13	13
Age of crown re-closure post-PCT (yr)	-	16	16
Number of large standing snags (trees/ha)	350	214	93

Table 5. Stand-level performance indices for density-manipulated upland black spruce plantations established at fixed IE levels subjected to PCT and CT treatments with genetic worth effects incorporated.

3.3 Extension of the model to address non-timber objectives

Conservation of biological diversity is the cornerstone of sustainable forest management (OMNR, 2005). Although the broader issues of forest-level structural complexity and connectivity, and overall wildlife habitat requirements were assumed to have been addressed at the landscape level during the forest management planning process, density management treatments are expected to affect biodiversity at both the stand and forest levels (Thompson et al, 2003). Specifically, at the stand-level, biodiversity would decline as a direct consequence of the reduction in structural complexity arising from IE, PCT and CT treatments, principally through the (1) establishment of monocultures, application of herbicides and species-specific thinning treatments which would reduce species diversity, (2) regulation of intertree spacing which would result in a decrease in spatial complexity, and (3) truncation of the diameter distribution due to the thinning-from-below treatment protocol which would reduce the degree of horizontal and vertical structural heterogeneity. Employment of improved planting stock would also result in a reduction in genetic diversity. Lastly, the lowering of the intensity of resource competition through IE, PCT and CT would result in a reduction in the rate of self-thinning and hence a decrease in the production of abiotic components (e.g., snags and coarse woody debris).

However, the question remains as to the degree of impact that a reduction in biodiversity arising from density management would have. In an extensive literature review of previous biodiversity impact studies augmented by model projections, Thompson et al., (2003) concluded that (1) the presence of large sturdy and standing snags was most important to the vertebrate population in terms of providing nesting and denning site, (2) the quality of coarse woody debris (CWD) in terms of its decay stage and size were more important than quantity in relation to providing cover, feeding areas, and den sites for wildlife, (3) having a devise vertical structure combined with the presence of fruiting species within the understory was most conducive to the songbird populations, and (4) canopy cover was important for many vertebrate species in regards to avoiding avian predators. However, it is evident that some of these requirements are specific to a given wildlife species and hence are inversely related (c.f., (3) and (4)). Consequently, achieving an optimal stand structure which complies with all the wildlife habitat requirements would be largely illusive. Thus regulating stand densities in order to realize biodiversity objectives will likely involve various tradeoffs.

The modular-based SSDMM can be used to provide direct or indirect structural metrics that address biodiversity objectives. For example, at any point in a stand's development the model provides estimates of horizontal and vertical structure (e.g., diameter and height distributions). Similarly, the degree of canopy closure and crown heights can be inferred from crown closure line or calculated from the live crown ratio isoline as presented in the SDMD graphic (Figures 2 and 3). Estimates of the approximate number, age and size of CWD components produced during stand development can be derived from the model using the density and total volume estimates (Newton, 2006). Once the threshold values for these biodiversity-based structural metrics are explicitly quantified, they could be added to the suite of performance measures. Hence, the SSDMM could be used to determine if a specific crop plan complied with not only volumetric, end-product, or economic objectives, but also biodiversity goals.

For example, consider the plantation scenarios but now with a CWD requirement superimposed. Although CWD requirement has yet to be defined in terms of absolute volumes, sizes and decay classes, it is evident that a CT treatment will remove a substantial amount of the larger-sized trees that would have naturally incurred mortality during the later stage of the rotation. In fact, relative to the control stand, 506 fewer merchantable-sized trees per hectare experienced mortality during the post-CT period. Hence this differential in CWD production is of concern given the importance of CWD to maintaining biodiversity. However, one approach in overcoming this CWD deficit is to leave more of the CT trees on site at the time of the treatment. Specifically, by changing the merchantability thresholds of the trees to be removed, more of the stem can be left behind on the forest floor. The minimum diameter of CWD has been defined as 7.5 cm in Ontario (OMNR, 2010) and hence by decreasing log length and increasing the minimum threshold diameters for both sawlogs and pulplogs, the residual amount of stem volume left on the site will increase.

To demonstrate, the third scenario was re-run with the following modifications: (1) all log lengths were set to 2.59 m; (2) the minimum diameter for pulplogs was increased from 10 to 12 cm; and (3) the diameter defining the merchantable top was set to 7.5 cm. Effectively, this increases the residual stem tip volume left behind given that this volume is defined as the volume between the top of the upper most log removed and the top of the merchantable stem. Table 6 lists a subset of the resultant yields and performance metrics for this modified regime relative to the previous PCT+CT regime where the CWD requirement was not explicitly addressed (Tables 4 and 5).

This comparison reveals that the production of large volumes of CWD via CT did result in a decline in merchantable volume productivity, economic efficiency and operability status. However, the treatment was profitable given that the revenue generated from the approximately 9 m³/ha of merchantable wood that was removed from the site exceeded the costs of acquiring and processing it. The CT treatment resulted in a substantial increase of relatively large CWD components (varying log lengths with diameters ranging from a minimum of 7.5 to a maximum of 12 cm). This CWD contribution should provide acceptable habitat to various wildlife species, particularly, the pine marten. Although not identical in terms of the volume of CWD produced, this scenario is similar to that proposed by Sturtevant et al. (1996) for pine marten habitat in western Newfoundland: i.e., providing old-growth stand structural attributes through the use of CT to generate downed CWD, which created denning and resting sites, subnivean access for cover, prey access, homeogeothermic regulation, and prey biomass (principally voles (genera *Microtus* and *Myodes*)), for the pine marten.

Index (unit)	Regime 3 – PCT+CT	Difference
Residual tip volume left on site at time of CT treatment (m³/ha)	24	+15
Total merchantable volume removed from the site via CT (m³/ha)	9	-30
Net Revenue arising from the CT treatment – SM ($K/ha)	0.1	-0.6
Net Revenue arising from the CT treatment – RLM ($K/ha)	0.4	-1.6
Relative land expectation value at rotation – SM (%)	74	-49
Relative land expectation value at rotation – RLM (%)	34	-38
Time to operability status (yr)	47	+2

Table 6. Subset of CT yield metrics and stand-level performance indices for upland black spruce plantations managed for the production of CWD.

3.4 Utility of SDMD-based decision-support models in forest management

SDMDs have an extensive history of development and use in forest management throughout many of the world's temperate and boreal forest regions. The SDMD developed by Ando (1962) for Japanese red pine (*Pinus densiflora* Siebold and Zucc.) in Japan was the first model to explicitly incorporate the reciprocal equations of the competition–density (C-D) and yield–density (Y-D) effect (Kira et al., 1953; Shinozaki & Kira, 1956) and the self-thinning rule (Yoda et al., 1963), into an integrated model framework. The reciprocal equation describes the relationship between mean tree size (C-D effect) or per unit area yield (Y-D effect) and density at specific stages of development within stands not incurring density-dependent mortality. The self-thinning rule describes the asymptotic relationship between mean tree size and density within stands undergoing density-dependent mortality. These core relationships were derived from empirical results and associated mathematical formulations arising from numerous plant competition experiments conducted during the 1950s and 1960s (e.g., Donald (1951), Kira et al. (1953), Hozumi et al., (1956), Shinozaki & Kira (1956), Holliday (1960), Yoda et al. (1963)). The SDMD is presented as a 2-dimensional bivariate graphic with density on the x-axis and mean volume on the y-axis upon which the reciprocal equations and self-thinning line are superimposed. Ando (1962) used the SDMD to design thinning schedules which would yield a specified quadratic mean diameter at rotation.

Following the successful introduction of the SDMD by Ando in 1962, Tadaki (1963) developed a SDMD for Sugi stands (*Cryptomeria japonica* D. Don.) in Japan and extended the utility of the model by illustrating how the reciprocal equation of the C-D effect could be used to estimate thinning yields. Later in 1968, Ando (1968) introduced a new set of SDMDs for Japanese red pine, Sugi, Hinoki cypress (*Chamaecyparis obtuse* (Siebold and Zucc.) Endl.) and Japanese larch (*Larix leptolepis* (Siebold and Zucc.) Gord.) stands in Japan. Using these new models, Ando demonstrated how they could be used as a decision-support tool in terms of evaluating the potential yield outcomes to various thinning treatments. In order to extend the applicability of the mean size – density relationship represented by the reciprocal equation of the C-D effect to stands incurring density-dependent mortality, Aiba (1975a,b) modified the Ando (1968) SDMD model for Sugi stands by replacing the reciprocal equation of the C-D effect with an empirical-based function where mean volume was expressed as function of both density and diameter.

Acknowledging the utility of the SDMD in forest management and silviculture decision-making, Drew and Flewelling (1979) introduced SDMDs to the forest management community in the Pacific Northwest through the development of a SDMD for coastal Douglas fir (*Pseudotsuga menziesii* (Mirb.) Franco.) stands. Since their introduction to the English-based forest science literature, numerous diagrams have been developed and utilized in stand-level management planning. These included SDMDs for Japanese red pine in Japan (Ando, 1962, 1968) and South Korea (Kim et al., 1987), Monterey pine (*Pinus radiata* D. Don.) in New Zealand (Drew & Flewelling, 1977) and Spain (Castedo-Dorado et al., 2009), Douglas fir in Spain (López-Sánchez & Rodríguez-Soallerio, 2009), lodgepole pine (*Pinus contorta* var. latifolia Engelm.) in the western USA (McCarter & Long, 1986; Smith & Long, 1987) and the Pacific Northwest (Flewelling & Drew, 1985), slash pine (*Pinus elliottii* Engelm. var. elliottii) and loblolly pine (*Pinus taeda* L.) in the southern USA (Dean & Jokela (1992) and Dean & Baldwin (1993), respectively), black spruce in the eastern and central Canada (Newton & Weetman, 1993, 1994), teak (*Tectona grandis* L.) in India (Kumar et al., 1995), pedunculate oak (*Quercus robur* L.) in Spain (Anta & González, 2005), Scots pine (*Pinus sylvestris* L.) and Austrian black pine (*Pinus nigra* Arn.) in Bulgaria (Stankova & Shibuya, 2007), Merkus pine (*Pinus merkusii* Jungh. et de Vriese) plantations in Indonesia (Heriansyah et al., 2009), and *Eucalyptus globulus* and *Eucalyptus nitens* short rotation plantations in Southwestern Europe (Pérez-Cruzado et al., 2011).

Analytically, the development of SDMDs has been characterized by a sequence of continuous incremental advancements in which increasingly complex and innovative model variants have been proposed. Acknowledging the paradigm shift in management focus from volumetric yield maximization to end-product recovery and value maximization (e.g., Barbour and Kellogg, 1990; Emmett 2006), and realizing the limitations of traditional SDMDs in addressing these new management objectives, the structural SDMD was introduced (Newton et al., 2004, 2005). Specifically, the structural model incorporated a parameter prediction equation system for recovering diameter distributions within the SDMD model architecture. More recently, an expanded version of the structural model was developed in order to address stand-level volumetric, end-product, economic and ecological objectives. To date, modular-based SSDMMs has been developed for jack pine (*Pinus banksiana* Lamb.) (natural-origin stands and plantations; Newton, 2009), black spruce and jack pine mixtures (natural-origin stands; Newton, 2011), upland black spruce (natural-origin stands and plantations; Newton, 2012a), and lowland black spruce (natural-origin stands; Newton, 2012b). These models were calibrated using extensive measurement data sets derived from hundreds of permanent and temporary sample plots situated throughout the central portion of the Canadian Boreal Forest Region. Consequently, the model and associated software suite (Croplanner) represents an operational and enterprise ready decision-support tool.

Essentially, these modular-based SSDMMs retain the ecological and empirical foundation of the original SDMD models, but in addition, incorporate estimation modules for predicting diameter, height, biomass, carbon, log, end-products and associated value distributions, and fibre quality attributes, at any point during a stand's development. The model allows managers to predict the consequences of a given crop plan in terms of realizing specified volumetric, end-product, economic or ecological objectives. In terms of its ability to forecast productivity, end-product and economic, the consequences of various density management treatments, the modular-based SSDMMs share a number of similarities to some of the existing stand-level density management decision-support models. Among others, these include SYLVER (Di Lucca, 1999) which was calibrated for Douglas fir and other coniferous species for use in western Canada, SILVA which was developed for Norway spruce (*Picea abies* (L.) Karst.) and other conifers and deciduous species for use in central Europe (Pretzsch et al., 2002), and MOTTI (Hynynen et al., 2005) which was developed for Scots pine and other conifers for use in Finland.

The SSDMM model architecture in which yield-density and allometric relationships provide the quantitative linkage among the component modules is readily adaptable in addressing new and evolving forest management objectives, as exemplified in the examples considered in this study. Given the large number of existing SDMDs combined with the transformative shift in management focus from volumetric yield maximization to product diversification, suggests that the modular-based SSDMM platform may have wide applicability in resource management.

4. Conclusion

The objectives of this study were to describe an enhanced stand-level decision-support model for managing upland black spruce stand-types, and demonstrate its operational utility in evaluating complex density management regimes involving IE, PCT and CT treatments. The traditional SDMD modeling approach along with its embedded ecological foundation is retained within the modular-based SSDMM structure. For a given density management regime, site quality, and cost profile, the model provides a broad array of yield metrics. These include indices of (1) overall productivity (mean annual volume, biomass and carbon increments), (2) volumetric yields (total and merchantable volumes per unit area), (3) log-product distributions (number of pulp and saw logs), (4) biomass production and carbon sequestration outcomes (oven-dried masses of above-ground components and associated carbon equivalents), (5) recoverable end-products and associated monetary values (volume and economic value of recovered chip and dimension lumber products) by sawmill-type (stud and randomized length), (6) economic efficiency (land expectation value), (7) duration of optimal site occupancy, (8) structural stability, (9) fibre attributes (wood density and branch diameter), and (10) operability status.

The utility of the model was exemplified by contrasting operationally relevant crop plans using a broad array of performance metrics. Specifically, the likelihood of (1) realizing an early operability objective via the use of PCT treatments within density-stressed natural-origin stand-types, and (2) enhancing end-product value through the use of PCT and CT within plantations, was evaluated. As demonstrated through these simulations, this ecologically-based model enables forest practitioners to rank alternative crop plans in order to select the most applicable one for a given objective. Additionally, the model provides annual and rotational estimates of volumetric, biomass and carbon yields, log distributions, recoverable products and monetary values, and fibre attributes, at both the diameter-class and stand levels. Although the results of these simulations are largely dependent on the input parameter settings (e.g., treatments (establishment densities, thinning treatments, site classes, rotation ages, product degrade values, variable and fixed cost profiles), the results readily illustrates the potential utility of the model in sustainable forest management.

The importance of the model in managing forest resources for the production high value solid wood products, bio-energy feed stocks, carbon credits, and ecosystem services including biodiversity, is explicitly acknowledged in the model's structure and output. Consequently, the model should be of utility as forest managers migrate to a value-added management proposition and attempt to address diverse objectives under varying constraints.

5. Acknowledgement

The author expressive his gratitude to the: (1) members of the participatory interagency advisory team, Dan Corbett, Northwest Science and Technology, Ontario Ministry of Natural Resources (OMNR), Jeff Leach, Tembec Inc, Ken Lennon, Northeast Science and Technology, Glen Niznowski, Regional Operations, OMNR, John Parton, Terrestrial Assessment Program, OMNR, Dr. Doug Reid, Centre for Northern Forest Ecosystem Research, OMNR, Dr. Mahadev Sharma, Ontario Forest Research Institute, OMNR, Al Stinson, Forestry Research Partnership (FRP) and Dr. Stan Vasiliauskas, Northeast Science and Technology, OMNR, for their constructive input and direction during the model calibration phase; (2) Daniel Kaminski, Natural Logic Inc. for assistance in the development of the VB.NET algorithmic version of the model; (3) Dave Wood and Staff of the Forest Ecosystem Boreal Science Coop for access to permanent and temporary sample plot data sets; and (4) Forestry Research Partnership and Canadian Wood Fibre Centre, for fiscal support.

6. References

Aiba, Y. (1975a). Effects of cultural system on the stand growth of Sugi-plantations (*Cryptomeria japonica*). II. A tendency of the constant in final stem volume yield of stands under actual stand density (in Japanese; English abstract). *Journal of the Japanese Forestry Society* (Tokyo, Japan), Vol. 57, No. 2, pp. 39-44, ISSN 1349-8509.

Aiba, Y. (1975b). Effects of cultural system on the stand growth of Sugi-plantations (*Cryptomeria japonica*). III. Estimate of the stem volume yield under actual stand density (H-D-p-V diagram) (in Japanese; English abstract). *Journal of the Japanese Forestry Society* (Tokyo, Japan), Vol. 57, No. 3, pp. 67-73, ISSN 1349-8509.

Ando T. (1962). Growth analysis on the natural stands of Japanese red pine (*Pinus densiflora* Sieb. et. Zucc.). II. Analysis of stand density and growth (in Japanese; English summary). Government of Japan, *Bulletin of the Government Forest Experiment Station* (Tokyo, Japan), No. 147.

Ando T. (1968). Ecological studies on the stand density control in even-aged pure stands (in Japanese; English summary). Government of Japan, *Bulletin of the Government Forest Experiment Station* (Tokyo, Japan), No. 210.

Anonymous. (2009). Tembec silviculture matrix, Supplementary Documentation (Section 22.0), 2009-2019 Forest Management Plan for the Romeo Malette Forest, Ontario Ministry of Natural Resources. Available from http://www.appefmp.mnr.gov.on.ca/eFMP/home.do?language=en (Accessed July 2011).

Anta, M.B. & González, J.G.Á. (2005). Development of a stand density management diagram for even-aged pedunculate oak stands and its use in designing thinning schedules. *Forestry*, Vol. 78, No. 3, pp. 209-216, ISSN 0015-752X.

Barbour R.J. & Kellogg R.M. (1990). Forest management and end-product quality: A Canadian perspective. *Canadian Journal of Forest Research*, Vol. 20, No. 4, pp. 405-414, ISSN 0045-5067.

Bell, W.F., Parton, J. Stocker, N., Joyce, D., Reid, D., Wester, M., Stinson, A. Kayahara, G. & Towill. B. (2008). Developing a silvicultural framework and definitions for use in forest management planning and practice. *Forestry Chronicle*, Vol. 84, No. 5, pp. 678-693, ISSN 0015-7546.

Carmean W.H., Hazenberg G. & Deschamps K.C. (2006) Polymorphic site index curves for black spruce and trembling aspen in northwest Ontario. *Forestry Chronicle*, Vol. 82, No. 2, pp. 231-242, ISSN 0015-7546.

Castedo-Dorado F., Crecente-Campo F., Álvarez-Álvarez P. & Barrio A.M. (2009). Development of a stand density management diagram for radiata pine stands including assessment of stand stability. *Forestry*, Vol. 82, No. 1, pp. 1-16, ISSN 0015-752X.

CCFM (Canadian Council of Forest Ministers) (2009). National forestry database. Available from http://nfdp.ccfm.org/silviculture/national_e.php.

Dean T.J. & Baldwin Jr. V.C. (1993). Using a density management diagram to develop thinning schedules for loblolly pine plantations. Government of the United States of America, Department of Agriculture, Forest Service, Southern Forest Experiment Station, New Orleans, Louisiana. *Research Paper* SO-275, 12 pp., ISSN 0149-9769

Dean, T.J. & Jokela, E.J. (1992). A density-management diagram for slash pine plantations in the lower coastal plain. *Southern Journal of Applied Forestry*, Vol. 16, No. 178-185, pp. 178-185, ISSN 0148-4419.

Di Lucca, C.M. (1999) TASS/SYLVER/TIPSY: systems for predicting the impact of silvicultural practices on yield, lumber value, economic return and other benefits. In: *Stand Density Management Planning and Implementation Conference*, C. Barnsey (Ed.), 7-16, ISSN 0824-2119, Edmonton, Alberta. Clear Lake Publishing Ltd., Edmonton, Alberta, Canada.

Donald, C.M. (1951). Competition among pasture plants. I. Intra-specific competition among annual pasture plants. *Australian Journal of Agriculture Research* Vol. 2, No. 4, pp. 355-376, ISSN 0004-9409.

Drew T.J. & Flewelling J.W. (1977). Some recent Japanese theories of yield-density relationships and their application to Monterey pine plantations. *Forest Science*, Vol. 23, No. 4, pp. 517-534, ISSN 0015-749X.

Drew T.J. & Flewelling J.W. (1979). Stand density management: an alternative approach and its application to Douglas-fir plantations. *Forest Science,* Vol. 25, No. 3, pp. 518-532, ISSN 0015-749X.

Emmett B. (2006). Increasing the value of our forest. *Forestry Chronicle,* Vol. 82, No. 1, pp. 3-4, ISSN 0015-7546.

Erdle T. (2000). Forest level effects of stand level treatments: using silviculture to control the AAC via the allowable cut effect. In: *Expert Workshop on the Impact of Intensive Forest Management on the Allowable Cut,* P.F. Newton, (Ed.), pp. 19-30, Canadian Ecology Centre, Mattawa (2000), Ontario, Canada. Available from http://www.forestresearch.ca/Projects/fibre/IFMandACE.pdf.

Flewelling J.W. & Drew T.J. (1985). A stand density management diagram for lodgepole pine. In: *Lodgepole pine: the species and its management,* D.M. Baumgarter, R.G. Krebill, J.T. Arnott, and G.F. Weetman (Eds.), pp. 239-244, ASIN B00139QVIW, Washington State University, Pullman, Washington, USA.

Heriansyah I., Bustomi S. & Kanazawa Y. (2009). Density effects and stand density management diagram for merkus pine in the humid tropics of Java, Indonesia. *Journal of Forestry Research,* Vol. 5, No. 2, pp. 91-113. ISSN 1993-0607.

Holliday, R. (1960). Plant population and crop yield. Field Crop Abstracts, Vol. 13, pp. 159-167, ISSN 0378-4290.

Hozumi, K., Asahira, T. & Kira, T. (1956). Intraspecific competition among higher plants. VI. Effects of some growth factors on the process of competition. *Journal of the Institute of Polytechnics* (Osaka City University, Japan), Series D, Vol. 7, pp. 15-34, ISSN 0305-7364.

Hynynen J., Ahtikoski A., Siitonen J., Sievanen R. & Liski J. (2005). Applying the MOTTI simulator to analyse the effects of alternative management schedules on timber and non-timber production. *Forest Ecology and Management,* Vol. 207, No. 1, pp. 5-18, ISSN 0378-1127.

Jack S.B. & Long J.N. (1996). Linkages between silviculture and ecology: an analysis of density management diagrams. *Forest Ecology and Management,* Vol. 86, No. 1-3, pp. 205–220, ISSN 0378-1127.

Kang K.Y., Zhang S.Y. & Mansfield S.D. (2004). The effects of initial spacing on wood density, fibre and pulp properties in jack pine (*Pinus banksiana* Lamb.). *Holzforschung,* Vol. 58, No. 8, pp. 455-463, ISSN 0018-3830.

Kim D.K., Kim J.W. Park S.K. Oh M.Y. &Yoo J.H. (1987). Growth analysis of natural pure young stand of red pine in Korea and study on the determination of reasonable density (in Korean; English abstract). Government of Korea, *Research Reports of the Forestry Institute* (Seoul, Korea), Vol. 34, pp. 32-40, ISSN 1225-0236.

Kira T., Ogawa H. & Sakazaki N. (1953). Intraspecific competition among higher plants. I. Competition-yield-density interrelationship in regularly dispersed populations. *Journal of the Institute of Polytechnics* (Osaka City University, Japan), Series D, Vol. 4, pp. 1-16, ISSN 0305-7364.

Kumar, B.M., Long, J.N. & Kumar, P. (1995). A density management diagram for teak plantations of Kerala in peninsular India. *Forest Ecology and Management,* Vol. 74, No. 1-3, pp. 125-131, ISSN 0378-1127.

López-Sánchez C. & Rodríguez-Soalleiro R. (2009). A density management diagram including stand stability and crown fire risk for *Pseudotsuga Menziesii* (Mirb.) Franco in Spain. *Mountain Research and Development*, Vol. 29, No. 2, pp. 169-176, ISSN 0276-4741.

Lindh B.C. & Muir P.S. (2004). Understory vegetation in young Douglas-fir forests: does thinning help restore old-growth composition. *Forest Ecology and Management*, Vol. 192, No. 2-3, pp. 285-296, ISSN 0378-1127.

McCarter J.B. & Long J.N. (1986). A lodgepole pine density management diagram. *Western Journal of Applied Forestry*, Vol. 1, No. 1, pp. 6-11, ISSN 0885-6095.

McKinnon L.M., Kayahara G.J. & White R.G. (2006). Biological framework for commercial thinning even-aged single-species stands of jack pine, white spruce, and black spruce in Ontario. Ontario Ministry of Natural Resources, Science and Information Branch, Northeast Science and Information Section. *Technical Report*, TR-046.

Newton P.F. (1997). Stand density management diagrams: review of their development and utility in stand-level management planning. *Forest Ecology and Management*, Vol. 98, No. 3, pp. 251-265, ISSN 0378-1127.

Newton P.F. (2006). Forest production model for upland black spruce stands – Optimal site occupancy levels for maximizing net production. *Ecological Modelling*, Vol. 190, No. 1-2, pp. 190–204, ISSN 0304-3800.

Newton P.F. (2009). Development of an integrated decision-support model for density management within jack pine stand-types. *Ecological Modelling*, Vol. 220, No. 23, pp. 3301-3324, ISSN 0304-3800.

Newton, P.F. (2011). *Development and Utility of an Ecological-based Decision-Support System for Managing Mixed Coniferous Forest Stands for Multiple Objectives*. In: *Ecological Modeling*, W-J. Zhang (Ed.),. pp. 300-361. Nova Science Publishers, Inc. New York, ISBN 978-1-61324-567-5.

Newton, P.F. (2012a). A decision-support system for density management within upland black spruce stand-types. *Environmental Modelling and Software*, Vol. 35, pp. 171-187

Newton, P.F. (2012b). A silvicultural decision-support algorithm for density regulation within peatland black spruce stands. *Computers and Electronics in Agriculture*, Vol. 80, pp. 115-125.

Newton P.F. & Weetman G.F. (1993). Stand density management diagrams and their utility in black spruce management. *Forestry Chronicle*, Vol. 69, No. 4, pp. 421-430, ISSN 0015-7546.

Newton P.F. & Weetman G.F. (1994). Stand density management diagram for managed black spruce stands. *Forestry Chronicle*, Vol. 70, No. 1, pp. 65-74, ISSN 0015-7546.

Newton P.F., Lei Y. & Zhang S.Y. (2004). A parameter recovery model for estimating black spruce diameter distributions within the context of a stand density management diagram. *Forestry Chronicle*, Vol. 80, No. 3, pp. 349-358, ISSN 0015-7546.

Newton P.F., Lei Y. & Zhang S.Y. (2005). Stand-level diameter distribution yield model for black spruce plantations. *Forest Ecology and Management*, Vol. 209, No. 3, pp. 181-192, ISSN 0378-1127.

Nilsen P. & Strand L.T. (2008). Thinning intensity effects on carbon and nitrogen stores and fluxes in a Norway spruce (Picea abies (L.) Karst.) stand after 33 years. *Forest Ecology and Management*, Vol. 256, No. 3, pp. 201–208, ISSN 0378-1127.

Ontario Ministry of Natural Resources (OMNR). (2005). Protecting what sustains us: Ontario's biodiversity strategy. OMNR, Queen's Printer, Toronto, Ontario, Canada. ISBN 0779479815.

Ontario Ministry of Natural Resources (OMNR). (2010). Forest management guide for conserving biodiversity at the stand and site scales. OMNR, Queen's Printer, Toronto, Ontario, Canada.

Pelletier G. & Pitt D.G. (2008). Silvicultural responses of two spruce plantations to midrotation commercial thinning in New Brunswick. *Canadian Journal of Forest Research*, Vol. 38, No. 4, pp. 851-867, ISSN 0045-5067.

Peltola A. (2009). Finnish Statistical Yearbook of Forestry. Available from http://www.metla.fi/julkaisut/metsatilastollinenvsk/index-en.htm.

Perez-Cruzado, C. Merino, A. & Rodriguez-Soalleiro, R. (2011). A management tool for estimating bioenergy production and carbon sequestration in Eucalyptus globulus and Eucalyptus nitens grown as short rotation woody crops in north-west Spain. *Biomass and Bioenergy*, Vol. 25, No. 7, pp. 2839-2851, ISSN 0961-9534.

Pretzsch, H. (2009). *Forest Dynamics, Growth and Yield*. Springer, ISBN 9783540883067. Verlag, Berlin and Heidelberg.

Pretzsch H., Biber P. & Dursky J. (2002). The single tree-based stand simulator SILVA: construction, application and evaluation. *Forest Ecology and Management*, Vol. 162, No. 1, pp. 3–21, ISSN 0378-1127.

Shinozaki K. & Kira, T. (1956). Intraspecific competition among higher plants. VII. Logistic theory of the C-D effect. *Journal of the Institute of Polytechnics* (Osaka City University, Japan), Series D, Vol. 12, pp. 69-82, ISSN 0305-7364.

Smith F.W. & Long J.N. (1987). Elk hiding and thermal cover guidelines in the context of lodgepole pine stand density. *Western Journal of Applied Forestry*, Vol. 2, No. 1, pp. 6-10, ISSN 0885-6095.

Stankova T.V. & Shibuya M. (2007). Stand density control diagrams for Scots pine and Austrian black pine plantations in Bulgaria. *New Forests*, Vol. 34, No. 2, pp. 123-141, ISSN 0169-4286.

Sturtevant B.R., Bissonette J.A. & Long J.N. (1996). Temporal and spatial dynamics of boreal forest structure in western Newfoundland: silvicultural implications for marten habitat management. *Forest Ecology and Management*, Vol. 87, No. 1-3, pp. 13-25, ISSN 0378-1127.

Tadaki, Y. (1963). The pre-estimating of stem yield based on the competition density effect (in Japanese; English summary). Government of Japan, *Bulletin of the Government Forest Experiment Station* (Tokyo, Japan) No. 154.

Thompson, I.D., Baker, J.A. & Ter-Mikaelian, M. (2003). A review of the long-term effects of post-harvest silviculture on vertebrate wildlife, and predictive models, with an emphasis on boreal forests in Ontario, Canada. *Forest Ecology and Management*, Vol. 177, No. 1-3, pp. 441-469, ISSN 0378-1127.

Tong Q.J., Zhang S.Y. & Thompson M. (2005). Evaluation of growth response, stand value and financial return for pre-commercially thinned jack pine stands in Northwestern Ontario. *Forest Ecology and Management*, Vol. 209, No. 3, pp. 225-235, ISSN 0378-1127.

Verschuyl J., Riffell S., Miller D. & Wigley T.B. (2011). Biodiversity response to intensive biomass production from forest thinning in North American forests – A meta-analysis. *Forest Ecology and Management,* Vol. 261, No. 2, pp. 221-232, ISSN 0378-1127.

Yoda K., Kira T., Ogawa H. & Hozumi K. (1963). Self-thinning in overcrowded pure stands under cultivated and natural conditions. *Journal of Biology,* (Osaka City University, Japan), Vol. 14, pp. 107-129, ISSN 0305-7364.

11

Application of Multi-Criteria Methods in Natural Resource Management – A Focus on Forestry

Mario Šporčić
University of Zagreb, Faculty of Forestry
Croatia

1. Introduction

Natural resource management refers to the management of natural resources such as land, water, soil, plants and animals which in accordance with the concept of sustainable development, a distinct emphasis puts on the way the management affects both present and future generations. In management and utilization of forests and forest land, as one of the most significant natural resources, the principle of the sustainable development is incorporated in a way that adheres to biological diversity, productivity, regeneration capacity, vitality and potential of the forests to fulfil, now and in the future, its important economical, ecological and social functions.

Forest resources and benefits that derive from them represent an important part in fulfilling the needs of humanity for energy, raw materials and quality of life. These benefits cover a broad range of goods and services. Among other, they include: wood, recreation, water, soil preservation, clean air, game, scenic beauty, etc. Many of such benefits and services can be simultaneously gained from a single forest stand. And even though many countries have legislative regulations that prescribe the course of forest management and/or protection of certain forest functions, there are still many debates on the issue how to manage forests and to which purposes. In general, we could say that today the basic postulate of forest management is multifunctional or multiple use of forests. It represents the manner in which the most of many different functions of forests are being utilized. In that sense, forest management should enable the most prudent usage of forests and forest land to provide some or all of respective products and services, while ensuring productivity and stability of forest ecosystems at the same time. In realizing these goals careful planning and decision making play a major role, and are considered to be especially significant for effective natural resource management and achieving the principles of sustainable development.

Planning and decision making in forest management represent a very complex task mainly because of the multitude and a broad spectre of criteria enrolled in the decision making process. That means that any decision making is under many different influences, and that at the same time every decision made affects many criteria of different nature. These influences and criteria encompass (Diaz-Balteiro & Romero 2008):

a. economical issues – wood production, non-wood forest products (forest trees fruits and flowers, seeds, mushrooms, honey, resin, humus) livestock, game management, hunting;
b. ecological and environmental issues – soil erosion, watershed regulation, biodiversity conservation, carbon sink, scenic beauty, influence on climate;

c.   social issues – recreational activities, tourism, level of employment, rural development, population settlement etc.

Moreover, the complexity of a large proportion of forestry issues is increasing due to the way in which different interest and social groups and organizations perceive the relative importance of specific criteria and appraise the management of forests, and thus judge the "quality" of forest resources management. The importance of specific criteria and evaluation of forest management in that sense depend on the personal standpoints and opinions of each individual i.e. group. Examples of such subjective assessments are often related to scenic beauty or recreational value of a certain forest area, or for example to game management and hunting. So, while someone preffers a specific game species and specific type of hunting, someone may want different kind of game and hunting, and someone else may be absolutely against hunting at all. Similar evaluations of forest management are related to the logging and creating certain revenue on the one hand and the protection and conservation of forests on the other hand.

All of the above mentioned daily increases the complexity of forest management, hinders the performance of forest operations and hardens the management conditions making the planning and decision making in forestry very demanding. And while in the past decision making and management in forestry have frequently been performed on the basis of common sense and/or past experiences, today's forestry with multiple criteria and functions calls for more flexible decision support. The complexity of today's business environment in forestry, the imperative of continuous increase of business and ecological efficiency, and multiple stakeholders with different interests impose the necessity to use new models and more precise methods. In that kind of a situation the joint use of multi-criteria decision making methods and different techniques of group decision making are becoming an important and potentially desirable way for solving forestry issues. It is considered that multi-criteria decision models and methods can provide to modern forestry, which has multiple aims and tasks, and multitude of interest groups with often conflicting interests, a strong and flexible support to decision making. Development and application of such methods that haven't traditionally been used in forestry could provide to management a new tool which can be a valuable aid both on strategic and operational level of decision making. The emphasis in doing so, is on the fact that decision proposals and decisions made must be based on the rational arguments.

This paper provides an overview of certain multi-criteria methods which can be used as a support for planning and decision making in forestry. Several methods of multiple-criteria decision making have been described and compared. Brief description and comparison presented in the paper includes following multi-criteria methods: Data Envelopment Analysis (DEA), Analytic Hierarchy Process (AHP), Multi-Attribute Utility Theory (MAUT), outranking methods, voting methods and Stochastic Multicriteria Acceptability Analysis (SMAA). The paper also gives a brief overview and analysis of problems and forest areas where multicriteria methods have been applied so far. The intention was to explain for which types of tasks and problems these methods can be applied in the field of forestry. That provides an insight into characteristics of the respective methods and a guideline to eventual choice of which method to apply. Many of the articles cited in the paper provide information on the existing experiences, reflect the actual role and significance of multi-criteria decision making in forestry and represent a valuable reference source that can be beneficial to students, researchers, experts and practitioners in forestry. The main aim of the paper is to raise the forestry profession's awareness about the importance and potential role that multi-criteria decision making can play in forestry. Concrete examples of the carried out investigations provide an insight into the possibilities, suitability and justification of the application of multi-criteria methods.

2. About decision making

Decision making process involves the choice of a specific solution among the set of different alternatives that solve a given problem. In a decision problem, there are goals to be achieved by the decision, the criteria used to measure the achievement of these objectives, the weights of those criteria that reflect their importance, and alternative solutions to a problem. Under the objective we consider the state of the system we want to reach by a decision, the criteria are the attributes that describe the alternatives and their purpose is to directly or indirectly provide information about the extent to which each alternative achieves the desired goal. In a given decision situation, the criteria are usually not equally important, and their relative importance is derived from the preferences of decision maker what is related to his system of values and other psychological characteristics. Data and information about these elements are with the appropriate actions summarized in one number for each alternative, and on the basis of these values the ranking of alternatives is determined. Figure 1 shows the basic procedures and steps in the process of decision making and problem solving.

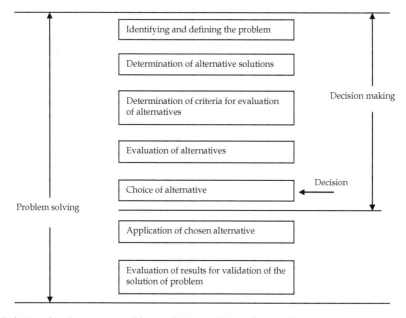

Fig. 1. Relationship between problem solving and decision making

Decision making is one of the major human activities, and one of the unavoidable tasks of managers. The decision situation is solved by adoption of a decision, which represents a selection of one action out of solutions available. The significance of decision making reflects in the fact that even if none of the possible solutions and actions have been chosen, the decision has been made - it has been decided not to choose or to do nothing.

3. Multi-criteria decision making approach

Multi-criteria decision making[1] falls within the wide range of operations research methods. As the name suggests, MCDM has been developed to enable analysis of multiple criteria situations and problems. It is usually applied in such cases where it is necessary to holistically consider and evaluate various decision alternatives, in which comprehensive analysis is particularly difficult due to a multiplicity of hardly comparable criteria and conflicting interests that influence the decision making process.

A number of MCDM methods have been developed, each of them with specific characteristics and different techniques that are applicable in appropriate circumstances and situations. For example, some methods are specially designed to manage risk and uncertainty, or for non-linear estimation, while others are focused on applications in conflict management tasks and objectives or on the use of incomplete or poor quality information. Many methods also come with a variety of settings and in different versions (eg, 'fuzzy' or stochastic versions, etc). Some are also slightly modified to better respond to tasks and problems in certain areas, including forestry. A detailed overview of operational research and multi-criteria decision making methods can be found in numerous sources (Vincke, 1992; Triantaphyllou, 2000; Koksalan & Zionts, 2001; Kahraman, 2008 etc).

The procedure of multi-criteria decision making involves the development of several alternatives that can no longer be improved by some criteria, while at the same time not ruined by the other criteria (Pareto optimality or efficiency). A comparison of selected alternatives is implemented considering all the previously set criteria and characteristics that influence the selection of a particular solution. As a result of a comprehensive comparison, the priority and rank of the observed alternatives is determined. In a group decision making individuals may, depending on their personal preferences, differently rank some alternatives. Comprehensive comparisons can also be made with assigning different weights to certain criteria, but also to opinions of individual participants. This includes the influence of different criteria and individual points of view which are taken into consideration together. In this way, MCDM methods can be used to analyze the situation of decision making and help in making the best possible or at least satisfactory decision.

Bearing in mind the above, it is considered that with the application of MCDM methods, many challenges in today's demanding and complex forest management planning can be facilitated and minimized. Many authors have written on that topic (Tarp & Helles, 1995; Krč, 1999, Kangas & Kangas, 2005; Herath & Prato, 2006 etc).

4. Main MCDM methods

This section gives a brief description of MCDM methods that can be applied in multifunctional forest management. Selected approaches represent different theories and schools as part of operational research. All presented methods have been tested and applied in forestry, and although many methods are not included in this paper, most of them are based on similar assumptions and theory as methods presented. For a more detailed study of specific methods and their application in forestry, relevant sources are given.

[1] Multiple Criteria Decision Making (MCDM) ili MCD Support (MCDS) ili MCD Aid (MCDA)

4.1 Data Envelopment Analysis (DEA)

In recent years DEA has become one of the central techniques in the analysis of productivity and efficiency. It was used in comparing organizations (Sheldon, 2003), companies (Galanopoulos et al., 2006), regions and countries (Vennesland, 2005). In determining business efficiency it was applied in banking (Davosir, 2006), education (Glass et al., 1999), agriculture (Bahovec & Neralić, 2001), wood industry (Balteiro-Diaz et al., 2006), forestry (Lebel, 1996; Kao, 1998; Bogetoft et al. 2003; Šporčić at al., 2008, 2009). DEA bibliography records more than 3200 papers published to 2001 (Tavares, 2002).

DEA is a methodology for determining the relative efficiency of production or non-production units (Decision Making Units, DMU) that have the same inputs and outputs, and vary according to the level of resources available and the activity levels within the transformation process. Based on the information about the actual inputs and outputs of all observed DMUs DEA constructs an empirical efficiency frontier and calculates the relative efficiency of each unit. The most successful units are those that determine the efficiency frontier, and the degree of inefficiency of other units is measured based on the distance of their input-output ratio in relation to the efficiency frontier.

While typical statistical methods are characterized as the central tendency approaches, which make their estimations based on the average production unit, DEA is based on extreme values and compares every DMU only with the best units. The basic assumption is that if some unit can produce Y outputs with X inputs, the other units should be able to do the same if they work efficiently. The center of the analysis lies in finding the 'best' virtual unit for every real unit. If the virtual unit is better than the original one, regardless if it achieves more outputs with the same inputs, or achieves the same outputs with less inputs, then the original unit is inefficient.

DEA relative efficiency scores are interesting to forestry experts, managers and researchers because of three DEA properties:

- direct comparison of units with multiple inputs and outputs with no need to know the explicit form of relation between inputs and outputs which can also be expressed in different units of measure,
- characterization of each organizational unit with one relative efficiency score,
- improvements which model suggests to inefficient units are based on actual results of organizational units that operate efficiently.

4.2 Analytic Hierarchy Process (AHP)

Analytic Hierarchy Process (AHP) is widely used and very popular method in many areas, including management of natural resources. Mendoza & Sprouse (1989), Murray & Gadow (1991), Kangas (1992) are among authors who have applied AHP in forestry, and the number of applications is steadily raising (Pykalainen et al., 1999; Ananda & Herath, 2003; Wolfslehner et al., 2005; Šegotić et al., 2003, 2007).

AHP has several advantages from the standpoint of multi-criteria and group planning. With the use of AHP, objective information, expert knowledge and subjective preferences can be considered jointly and simultaneously. It can also take into consideration qualitative criteria, while other methods usually require quantitative values for the selection of the alternatives. Solving a complex decision problems using this method is based on their decomposition into components: goal, criteria (sub-criteria) and alternatives. These elements are then taken into a multi-level model (hierarchical structure) where the goal is on the top, and the main criteria represent the first lower level. The criteria can be broken down into sub-criteria, and on the lowest level of hierarhical structure, there are alternatives. Another important component of the method is a mathematical model which calculates the priorities (weights) of the elements on the same level of hierarchical structure. The method is based on comparisons of pairs of alternatives, each one with the other, while expressing the intensity and weight preferences of one alternative over another. The criteria are compared in the same way, whereby preferences are expressed by using Saaty's scale (Saaty 1977, 1980).

Negative aspect of the method is that it does not allow any reluctance and hesitation in the comparisons. In the management of natural resources, much of the information and data underpinning the planning and decision making is characterized by a certain level of insecurity and uncertainty. Furthermore, the number of comparisons significantly increases with the number of alternatives and criteria, which can be expensive and demanding. To overcome these limitations different AHP models have been developed. A'WOT combines AHP and well-known SWOT analysis (Kurttila et al., 2000), Analytic Network Process (ANP) is an extention of AHP (Satya, 2001) etc. Such hybrid models also have the same basic idea of pair-wise comparisons as practical, pedagogical and intuitive approach. Popularity of the method is primarily based on the fact that it is very close to the way in which individual intuitively solves complex problems by dismantling them into simpler ones.

4.3 Multi-Attribute Utility Theory (MAUT)

MAUT is a structured decision-making procedure for making a selection among different alternatives in relation to fulfilling a selected criteria. It is based on the utility theory that systematically seeks to validate and quantify the user's choice, usually on a scale 0-1 (Keeney & Raiff, 1976). Based on MAUT methodology there have been developed methods such as HERO and SMART, which rank given alternatives directly by assigning them numerical values proportional to their importance (Venter et al. 1998; Kajanus et al., 2004).

Simple Multi-Attribute Rating Technique (SMART) was developed in the early 1970s within the multiattribute utility theory. SMART methodology has many similarities with the basic idea of AHP method, but the main difference is that SMART does not use the comparison in pairs. Instead, the ranking of alternatives is carried out directly. Direct ranking means that the criteria are directly assigned numerical values proportional to their importance. Accordingly, alternatives are assessed with respect to each decision criterion by simply giving them relative numerical values that reflect their priorities. Most often, after the selection of criteria, the main criterion is determined and given a value 100. All other criteria are assigned values between 0 and 100, depending on their importance to the main criterion. According to the same principle each alternative is assigned a certain value in relation to individual criteria. The best alternative is given the value 100, while all other alternatives have values between 0 and 100 depicting their rank. When the importance of certain criteria and priorities among alternatives have been identified, SMART uses the same computational procedures as AHP. Examples of using SMART in natural resource management include Venter et al. (1998), Kajanus et al. (2004), etc.

4.4 Outranking methods

Outranking methods represent European or French School of MCDM. Many different methods have been developed, and among them PROMETHEE and ELECTRE have been applied in forestry (Kangas et al., 2001). These methods compare the alternatives in pairs, on the basis of so-called pseudo-criteria. Pseudo-criteria are two threshold values, the indifference threshold and preference threshold, which describe the difference in the severity of preferences between two alternatives. If the difference is less than the indifference threshold, the alternatives are considered to be indifferent in regard to that criterion. If the difference exceeds the preference threshold, better alternative is considered to be better without a doubt. If the difference is larger than the indifference threshold, but less than the preference threshold, priority between alternatives is uncertain.

Calculations are carried out in different ways in PROMETHEE and ELECTRE, and both methods have several versions to suit different situations. The main advantage of these methods is that they do not require as complete preference data as AHP. Disadvantage is that these are fairly obscure methods that are quite difficult to understand and interpret.

4.5 Voting methods

Voting is a familiar way of expressing opinions and influencing important matters. Voting techniques can be applied in MCDM when determining the criteria. The criterion that gets the most votes is considered the most important. Another example might be a vote on the suitability of alternatives with respect to certain criteria. Voting can be conducted under the principle "one man, one vote" or by giving a voter a certain number of votes. In Approval Voting voter gives a vote to each option deemed acceptable. In so-called Borda Count each voter gives n votes for the best option, $n - 1$ votes for the next, and so on until one vote remains for the worst option. These methods are some examples of many voting techniques. Voting techniques have been developed to handle situations with the low quality of data on preferences. Simplicity and comprehensiveness of the voting techniques are their main advantage, especially in group planning and decision making. By including more information, they increasingly resemble to SMART method. The general attitude is that voting methods should not be modified and further complicated for applications for which there already exist specific multi-criteria methods. The Multi-criteria Approval method is based on approval voting and has been applied in forestry (Laukkanen et al., 2002, Kangas & Kangas 2004). Shields et al. (1999) and Hiltunen et al. (2008) also applied voting methods.

4.6 Stochastic Multicriteria Acceptability Analysis (SMAA)

Similar to SMART, SMAA actually represents a set of methods. They were originally developed for discrete multi-criteria problems with uncertain or inaccurate criteria data, and where, for some reason, it was not possible to obtain data on weights and preferences from the decision makers. SMAA methods are based on determining the weight values that would make each alternative the preferred one, or that would give a certain rank to an alternative. Key indicators of SMAA include so-called acceptability indices, which describe the probability of placing an alternative in a certain rank. If the weight values of the criteria are not predetermined, the acceptability indices show the dominance of alternatives among all possible weighting combinations. The overall acceptability index can be calculated as a weighted average of the probable alternative ranks, with the most weight for the first place, then second and so on. This method is close to forest management where, due to a strong uncertainty in the planning, usually none of the alternatives under consideration can be safely declared as the best one.

The first applications of SMAA methods in forestry have been implemented in the context of ecosystem management planning (Kangas et al., 2003, Kangas & Kangas, 2004). Since SMAA includes many useful characteristics it is increasingly gaining interest in today's forestry and natural resources management (Kangas et al. 2006; Diaz-Balteiro & Romero, 2008).

4.7 Comparison of MCDM methods

Presented methods significantly differ one from another. Neither of reviewed methods is universal or the best, not even applicable in all cases. In fact, to a different situations and problems best suite different methods. Selection of appropriate method requires knowledge of various methods, their preferences, strengths and limitations as well as the characteristics and requirements of specific situation and problem in planning and decision making. Table 1 shows the comparison of presented and some additional MCDM methods.

MCDM method	Cost of implementation	Data reqiurement	Ease of sensitivity	Economic rigor	Management under-standing	Mathematical complexity	Parameter mixing-flexibility
DEA	M	M	L	M	L	H	M
AHP	M	M	L	L	M	L	H
Expert systems	H	H	L	H	M	H	H
Goal program	M	M	M	H	L	H	L
MAUT	H	H	M	M	M	M	H
Outranking	M	M	L	M	L	M	M
Simulation	H	H	H	H	H	H	M
Scoring models	L	L	L	L	H	L	H

H- high; M - medium; L - low

Table 1. MCDM methods' characteristics (Sarkis & Weinrach, 2001)

Table 1 shows that none of the methods does not dominate over the other methods. For example, when compared to other methods DEA is moderately demanding regarding costs of implementation and data collection. Sensitivity to changes in data is small, and the managerial understanding of the method is relatively low, mainly due to its mathematical complexity. The results are easy to interpret because it ranks compared units by their efficiency while flexibility allows including more parameters and factors in the analysis.

It is generally difficult to directly compare different methods. Each method has its advantages and disadvantages. The application often depends on the decision environment, where the availability of data, time and costs influence the selection of specific method. In any case, when applying in analysis researchers, experts and managers should be aware of their characteristics, both advantages and limitations.

5. Applications of MCDM in forestry

Although MCDM has been present in forestry for more than 30 years (Field, 1973), some newer approaches and techniques of multi-criteria and group decision making have become more significant in the early 1990s (e.g. Kangas, 1992). In that time period, a significant number of papers dealing with various problems of forestry have been published. This section will present some examples of conducted investigations and MCDM applications in certain forestry areas. Conditionally determined areas of forestry in which MCDM methods have been applied so far can be defined as follows (Diaz-Balteiro & Romero, 2008):

- Harvesting
- Extended harvesting
- Forest biodiversity
- Forest sustainability
- Forestation
- Regional planning
- Forestry industry
- Risk and uncertainty

Forest harvesting and it's planning is the first forestry area in which MCDM paradigm has been widely applied (Kao & Brodie, 1979; Hallefjord et al., 1986). Howard & Nelson (1993) used MAUT methods for solving a specific problem of forest harvesting. Diaz-Balteiro & Romero (1998) used AHP in planning of forest harvesting, while Heinonen & Pukkala (2004) in harvest scheduling issues used a version of HERO method.

Extended forest harvesting besides timber and logging, includes the problems of non-wood forest products. Thus, Arp & Lavigne (1982) in proposed multi-criteria model included timber, recreation, hunting and wildlife. Hyberg (1987) set a MAUT model with two attributes: production of wood and aesthetic values. Rauscher et al. (2000) with regard to more non-wood criteria, evaluated four management alternatives using AHP. Laukkanen et al. (2002, 2005) applied different voting techniques to several problems of forest exploitation in Finland. Kangas et al. (2005) used SMAA method with recreational and environmental criteria, and Pauwels et al. (2007) compared several silvicultural alternatives of Larix stands with the use of ELECTRE.

The field of forest biodiversity has been, from the position of MCDM, associated with the management of national parks, reserves, etc., where the selection of management activities leads to application of different MCDM methods. For example, Kangas (1994) applied AHP in the management of protected natural areas in Finland. Rothley (1999) used MCDM methods for designing optimal biodiversity network in Canada. Kurttila et al. (2006) used MAUT to find the optimal compensation for forest owners who orient their forest management towards biodiversity conservation.

Efforts to connect issues of forest sustainability with MCDM approach are relatively new. Its applications in this area are mainly related to the assessment of management quality based on the analysis and aggregating of different sustainability indicators in a single index as an overall measure of forest systems sustainability (Mendoza & Prabhu, 2003; Manessi & Farrell 2004). Kant & Lee (2004) used voting techniques and Borda method for the evaluation and ranking of forest management plans with regard to sustainability. In a similar problem Mendoza & Dalton (2005) used AHP, and Huth et al. (2005) PROMETHEE.

In the area of re/af/forestation the first MCDM paper was published by Walker (1985) who developed a methodology for reforestation planning, taking into account several species, silvicultural treatments, etc. More authors combined MAUT and AHP in their approach to this issue (Kangas, 1993; Nousiainen et al., 1998). Liu et al. (1998) used AHP to assess regional forestation projects in China. Giliams et al. (2005) compared AHP, ELECTRE and PROMETHEE in choosing the best afforestation alternative in Belgium.

In the field of regional planning MCDM methods are represented in papers which deal with planning and efficiency of forest management practice in certain national or regional area (Buongiorno & Svanquist, 1982; Faith et al., 1996; Liu et al., 1998). Kangas et al. (2001) analyzed a forest management case in eastern Finland by three multi-criteria techniques:

MAUT, ELECTRE and PROMETHEE. By applying DEA method Kao (1998) measured the efficiency of forest districts in Taiwan, Vennesland (2005) measured the efficiency of subsidies in supporting regional development in Norway and Hiltunen et al. (2008) used five voting methods in strategic forest planning in state forests of Finland.

Considering forestry industry, most papers are related to efficiency evaluation with the use of DEA methods. For example, Yin (1998) analyzed efficiency of 44 paper companies in United States, Nyrud & Bergseng (2002) measured the efficiency of 200 sawmills in Norway, Sowlati & Vahid (2006) evaluated efficiency of Canadian wood product industry, Diaz-Balteiro et al. (2006) analyzed efficiency and innovation activities in Spanish wood industry.

Risk and uncertainty are strongly present in forest management where incomplete data and insufficient information in planning and decision-making often do not allow more accurate assessments and plans. MAUT techniques are therefore most widely used MCDM approach to problem of risk and uncertainty (Pukkala, 1998; Lexer et al., 2000; Ananda & Herath, 2005). Leskinen et al. (2006) used AHP to evaluate the uncertainty associated with the preferences of forest owners in Finland, Kangas (2006) used SMAA for analyzing risks in an actual decision making process.

Cited papers are just some examples of the conducted investigations. The number of MCDM papers in forestry has evolved significantly in the last years. Some authors give a survey of multi-criteria applications in forestry and list more than 250 papers published in major English language journals in the last 30 years (figure 2).

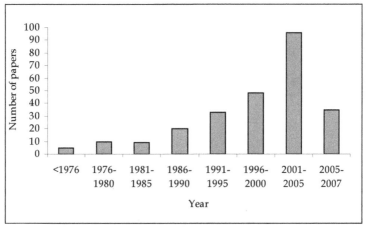

Fig. 2. Number of published multi-criteria papers in forestry (Diaz-Balteiro & Romero, 2008)

The literature also shows that MCDM methods have been applied to a wide range of forestry issues. The main forestry topics in which MCDM methods have been applied could be roughly categorized as already stated to: harvesting; extended harvesting; biodiversity; sustainability, etc. The classification itself cannot be precise because some papers can be divided into several areas or they use more than one method. Still, overview of published papers provides information on the investigated problems and applied MCDM methods in forestry. Figure 3 gives the number of multi-criteria papers in different forestry topics.

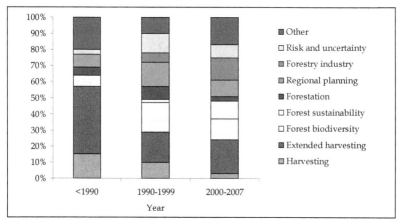

Fig. 3. Shares of MCDM papers in different forestry topics (Diaz-Balteiro & Romero, 2008)

6. Some examples of conducted investigations

This section gives more detailed overview of two investigations where MCDM approach was applied in forestry. Investigations were carried out within research projects and the needs of the state forestry company in the Republic of Croatia. One is related to biological parameters in the evaluation of natural resources (Posavec, 2005), and the other to efficiency of organizational units in forestry (Šporčić, 2007). The presented examples will point out the justification, and the applicability of multi-criteria methods in forestry.

6.1 Selection of biological parameters in the evaluation of natural resources

This research identified the values and value principles applied in evaluation of natural resources. The processed data are related to the Forest Management Unit "Gaj" of the Forest Administration Našice, Croatia. Using the potential method and the eigenvector method, the biological parameters that participate in the calculation of the total value of the natural forest resources were analysed. The adopted premise was that current methods have not been sufficiently exact, so that the new dynamic model should be used for the determination of the forest value. Conducted analysis and the development of new dynamic model included the application of AHP method.

The basic objective was to set up a scientific approach to evaluating a forest resources and establish a model applicable in practice. Parameters needed for forest value assessment, were evaluated by the experts (decision makers) from the field of forestry (Faculty of Forestry, University of Zagreb). Not all of the parameters in the evaluation had the equal weights. To the decision makers, a "verbal scale" for priority expression of one alternative related to another was available. In re-calculation of these verbal priorities into numerical ones, one of the twenty-seven most often used scales was used, as described in Saaty T.L. (1980). The following verbally expressed priorities were considered: Indifference; Moderate Priority; High Priority; Significant Priority; Absolute Priority, and their intermediate degrees, if a decision maker needed them in expressing priorities. Group decision as a potential method consisted of each group member defining their hierarchy, and a consensus at an alternative level (Čaklović et al., 2001). Thereby a group preference graph was made as a "sum" of individual preference graphs, followed by a group potential. This makes sense particularly if the decision makers do not agree with the criterion choice. Another reason for using the model of group decision was the possibility of measuring the distances among decisions of group members. If group members have coinciding opinions on alternative ranking, there is no need to insist on adjusting the standpoints related to the criterion choice. The comparison by pairs was based on Analytical Hierarchy Process (AHP). The method is supplied by the programme Expert Choice that helps in decision-making on complex issues with several criteria and possible actions. It is designed to model our way of thinking and simplify the process of decision-making (Šegotić, 2001).

In order to be used in a dynamic model for determining the value of selected management unit, the calculated values have been classified in four basic management aims as presented in Figure 4: economic target, management, direct use and indirect use.

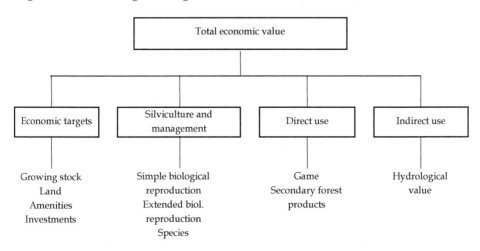

Fig. 4. Hypothetical criteria and parameters to be used in decision-making

Accordingly, forest value is the function of the economical, silvicultural/management, direct and indirect values expressed by the formula:

$$Vf = f (Ve + Vu + Vd + Vi)$$

Ve = economic value (Vgs + Vl + Va + Vi)
Vu = the value of silviculture and management (Vsbr + Vebr + Vspecies)
Vd = direct value (Vg + Vsfp)
Vi = indirect value (Vh + Vnwfp)

The presented aims and parameters of management represent parts of the common formula for determining the total value (Posavec, 2001). The total value of the management unit was presented through the total sum of the parameters and their weights (w):

$$Vt = (w1\ Vgs)+(w2\ Vl)+(w3Vsbr)+ (w4\ Vebr)+ (w5Vsfp)+ (w6Vg)+ (w7\ Vh)+ (w8Va)+ \\ (w9Vi)+ (w10\ Vnwff)+ (w11\ Vspecies)$$

Vt   = total forest value
Vgs   =growing stock value

Vl   = land value $(\sum_{i=1}^{i=n} V_z)$

Va   = amenities value (reduced by amortisation)
Vi   = investments value
Vsbr   = value of simple biological reproduction
Vebr   = value of extended biological reproduction
Vg   = value of hunting management

Vsfp   = value of secondary forest products $(\sum_{i=1}^{i=n} V_{sp})$

Vh   = hydrological value
Vnwff   = value of non-wood forest functions
Vspecies   = value of managed dominant forest species
In table 2, the results are the ranks of all parameters that contribute to forest value and were obtained by potential method. Variable X is the potential value of each parameter. The angle of 18.60 degrees is the measure of inconsistency within the allowed limits. One significant detail is that angle as a measure of group inconsistency does not have any impacts, although the programme displays it. It is significant to measure mutual distances between group members in terms of differences in individual preferences. The obtained distances make up the distance matrix as the basis for the clustering of the group. The sums of total weights form value 1, while individual parameters are presented by their size, which means that the highest priority in this case is the one of non-commercial forest functions.

			- Group ranking -			
			Forest value			
			Members			
Person 1	Person 2	Person 3	Person 4	Person 5	Person 6	Person 7
0.143	0.143	0.143	0.143	0.143	0.143	0.143

showWeights: groupAim   base = 2   Options = weight

Level 2:   alternatives
Comp_1

Weight = 1.000   InvInc = 0.337   (Angle = 18.60 deg)

Nodes:
nwff 0.121   (X= 0.453)
species 0.119   (X= 0.434)
vsbr 0.106   (X= 0.265)
vgs 0.103   (X= 0.230)
game 0.096   (X= 0.117)
vebr 0.090   (X= 0.022)
vsfp 0.088   (X= -0.008)
vl 0.085   (X= -0.059)
hv 0.082   (X= -0.101)
va 0.059   (X= -0.582)
vi 0.052   (X= -0.770)
Total weight = 1.000

Table 2. Group ranking of parameters by potential method

The AHP model in this case had a very simple structure (according to Figure 4). All parameters were alternatives, and were used for calculating total forest value. Supported by the eigenvector method, an attempt was made to obtain their weights. The basis for calculating the weights were estimates of the experts who carried out the comparisons per pairs of all given parameters. Supported by the programme Expert Choice, five rank lists with parameter weights for calculating forest value were made. An example of one expert's results is shown in Figure 5.

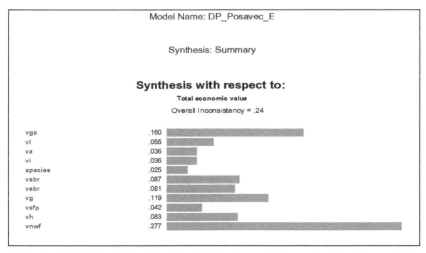

Fig. 5. Parameter rank list of the expert Posavec

If there are additional requirements for individual ranks (i.e. the feeling for forest value), a single rank can be adjusted to given reasons. A special programme can also calculate total forest value independently. In using the potential method, a constant exponential base is set (base value = 2). By changing this base, only relations between ranks can be changed, not their order. The total value of the management unit as calculated by the potential method amounted to 512,301,542.17 kunas (1 eur = 7.4 kunas). The total value calculated by the eigenvector method was 779,716,802.70 kunas. The difference between these two methods, depending on the estimate and parameter ranking, gave a value of 267,415,260.53 kunas. This result shows that a small difference in the size of the ranked parameter results in a great difference in final data. This relates particularly to calculations of the highly estimated non-wood forest values, which have the strongest impact on the final result.

The selected methods are based on pair comparison. Such comparison results in the development of a preference graph, while the number of comparisons per pairs grows in dependence of the given model. The advantage of the analysed dynamic model is obvious due to a decrease in input data. Another advantage of the analysed models is the possibility of clustering of particular groups, i.e. the measurements of the distances between the individual members and their preferences. The disadvantage of these methods is seen in subjective decisions made by individual experts (Posavec et al., 2006).

The developed dynamic models consider the characteristics of forest potentials, and follow the dynamics of the developing conditions within a forest stand, supporting the models of sustainable forest management. The method supports modern evaluation trends in forestry, using available computer program and multi-criteria methods in developing dynamic models for evaluation of forest resources' value.

6.2 Measuring efficiency of organizational units in forestry by nonparametric model

This research assesses the efficiency of basic organizational units in the Croatian forestry, forest offices, by applying Data Envelopment Analysis (DEA). Determination of efficiency is becoming increasingly important in many areas of human activity. Approach to this problem is particularly interesting when there are no clear success parameters, and when the efficiency of using several different resources/inputs is measured for achieving several different outputs. In forestry, the determination of efficiency of forestry companies is extremely complex because of multiple goals of forest management, i.e. its multiple inputs and outputs. In such conditions, the right evaluation method must be used in order to determine whether the resources are used efficiently.

The research included 48 forest offices. The selected forest offices were the representatives of four main regions in Croatian forestry: lowland flood-prone forests (I), hilly forests of the central part (II), mountainous forests (III) and karst/Mediterranean forests (IV). Each region was represented by two forest administrations i.e. by six forest offices from each forest administration (figure 6).

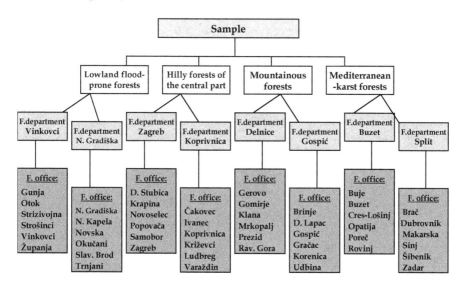

Fig. 6. Sample of the forestry organizational units involved in the research

The relative efficiency of compared forest offices was calculated with the most frequently used DEA models - CCR (Charnes-Cooper-Rhodes) and BCC (Banker-Charnes-Cooper) model. Since DEA was introduced by Charner, Cooper and Rhodes (Charnes et al., 1978) several analytical models have been developed depending on the assumptions underlying the approach. For instance, the orientation of the analysis toward inputs or outputs, the existence of constant or variable (increasing or decreasing) returns to scale and the possibility of controlling inputs. According to Farrell (1957), technical efficiency represents the ability of a decision making unit (DMU) to produce maximum output given a set of inputs and technology (output oriented) or, alternatively, to achieve maximum feasible reductions in input quantities while maintaining its current levels of outputs (input oriented). In this study, output oriented DEA was used, given it is more reasonable to argue that forest area, growing stock and other inputs should not be decreased. Instead, the goal should be increased outputs of forest management, and improved general state of forests. For computing the applied models, DEA Excel Solver software was used.

Given the selected orientation and the diversity of units characterizing the example, CCR model with constant returns to scale was applied first. Following Cooper et al. (2003), analysis began by the commonly used measure of efficiency (output/input ratio) and an attempt to find out the correponding weights by using linear programming. To determine the efficiency of n units (forest offices) n linear programming problems must be solved to obtain the value of weights (v_i) associated with inputs (x_i), as well as the value of weights (u_r) associated with the outputs (y_r). Assuming m inputs and s outputs and transforming the fractional programming model into a linear programming model, the CCR model can be formulated as (Cooper et al., 2003):

Max $$\theta = u_1 y_{10} + \dots + u_s y_{s0}$$

Subject to: $$v_1 x_{10} + \dots + v_m x_{m0} = 1$$

$$u_1 y_{1j} + \dots + u_s y_{sj} - v_1 x_{1j} - \dots - v_m x_{mj} \leq 0 \quad (j = 1, 2, \dots, n)$$

$$v_1, v_2, \dots, v_m \geq 0$$

$$u_1, u_2, \dots, \mu_s \geq 0 \tag{1}$$

Due to lack of information concerning the form of the efficiency frontier, an extension of CCR model, BCC model was also used. This model incorporates the property of variable returns to scale. The basic formulation of the model is as follows:

Max $$\theta = u_1 y_{10} + \dots + u_s y_{s0} - u_0$$

Subject to: $$v_1 x_{10} + \dots + v_m x_{m0} = 1$$

$$u_1 y_{1j} + \dots + u_s y_{sj} - v_1 x_{1j} - \dots - v_m x_{mj} - u_0 \leq 0 \quad (j = 1, 2, \dots, n)$$

$$v_1, v_2, \dots, v_m \geq 0$$

$$u_1, \mu_2, \dots, u_s \geq 0 \tag{2}$$

Where u_0 is the variable allowing identification of the nature of the returns to scale. This model does not predetermine if the value of this variable is positive (increasing returns) or is negative (decreasing returns). The formulation of the output oriented models can be derived directly from models described in (1) and (2) (Cooper et al., 2003).

The identification of inputs and outputs is, besides the choice of the basic model, considered to be the only element of subjectivity in DEA. They were selected so as to reflect business activities of the investigated DMUs – forest offices as the basic organisational units of the Croatian forestry, which perform the basic professional and technical operations in forest management and where the most income and direct costs incur.

As inputs the model included:

- Land, I1 – forest area in thousand hectares
- Growing stock, I2 – volume of forest stock in cubic meters per hectare
- Expenditures, I3 – money spent in hundred-thousand croatian kunas (7,4 kn ≈ 1 EUR)
- Labour, I4 – number of employees in persons

As outputs the model included:

- Revenues, O1 - yearly income in hundred-thousand croatian kunas (7,4 kn ≈ 1 EUR)
- Timber production, O2 – timber harvested in cubic meters per hectare
- Investments in infrastructure, O3 – forest roads built in kilometres
- Biological renewal of forests, O4 – area of conducted silvicultural and protection works in hectares

Table 3 presents the descriptive statistics of the variables used in the analysis. A wide variation in both inputs and outputs is noticeable. Such variation is not unexpected, since the sample involves all representative areas managed by Croatian forests Ltd. However, it may also be a sign of bad management of resources in individual forest offices.

Variable	Mean	St. deviation	Min	Max	Total
Inputs					
Area, 10^3 ha	11.42	10.36	2.60	49.87	547.96
G. stock, m^3/ha	214.98	91.94	51.85	418.00	-
Costs, 10^5 kn	152.35	93.61	23.24	470.31	7312.99
Employees, N	42	21	8	100	2.007
Outputs					
Income, 10^5 kn	157.20	106.40	21.12	538.41	7545.68
Harvest, m^3/ha	3.06	2.19	0.00	8.78	-
Investments, km	2.24	4.29	0.00	22.59	107.48
B. renewal, ha	422.26	606.34	30.21	3846.34	20268.47

Table 3. Descriptive statistics of the variables used in the DEA model

Technical efficiency was determined individually for each forest office. The average CCR efficiency of the investigated forest offices was 0.829, which means that an average (assumed) forest office should only use 82.9% of the currently used quantity of inputs and produce the same quantity of the currently produced outputs, if it wishes to do business at the efficiency frontier. In other words, this average organisational unit, if it wishes to do business efficiently, should produce 20.6%[2] more output with the same input level. According to the BCC model, the average efficiency is 0.904. This means that an average forest office should only use 90.4% of the current input and produce the same quantity of output, if it wishes to be efficient. In other words, to be BCC efficient it should produce 10.6%[3] more outputs with the same inputs. Scale efficiency (ratio between CCR and BCC scores) shows how close or far the size of the observed unit is from its optimal size. The scale efficiency of 0.919 means that the analysed forest offices would increase their relative efficiency on average by 8% if they adapted their size or volume of activities to the optimal value. The main results obtained by the output-oriented DEA are given in table 4.

---

[2] It can be easily obtained that $20.6\% = (1 - 0.829)/0.829$
[3] It can be easily obtained that $10.6\% = (1 - 0.904)/0.904$

	CCR model	BCC model	Scale efficiency
Number of forest offices (DMUs)	48	48	48
Relatively efficient DMUs	15	24	16
Relatively efficient DMUs (in %)	31 %	50 %	33 %
Average relative efficiency, E	0.829	0.904	0.919
Maximum	1,000	1,000	1,000
Minimum	0.407	0.524	0.501
Standard deviation	1.170	0.129	0.138
DMUs with efficiency lower than E	23	18	12

Table 4. Results obtained with the base case DEA models

Based on the efficiency results of forest offices grouped according to their structural characteristics (surface area, growing stock, number of employees), it has been determined that the highest levels of efficiency were recorded for forest offices that manage 10 to 15.000 hectares, and for the forest offices with growing stock ranging between 200 and 300 m³/ha i.e. over 300 m³/ha. It has been also determined that the highest level of efficiency is achieved by forest offices with the highest number of employees, and that forest offices in the region of flood-prone forests have the highest efficiency scores.

DEA method gives to management the possibility to rank organizational units based on analysis and comparisons of their relative efficiency. For inefficient units the projections on the efficiency frontier and the sources of inefficiency are determined. In this way, potential changes in inputs/outputs required to achieve technical efficiency are determined, and the objectives which inefficient units should fulfil in order to become efficient are recognized.

7. Conclusion

In the last twenty years or so, the general framework of forest management has changed dramatically. Multiple goals are today typical for forestry. Forest management has to produce a certain revenues while at the same time it should promote protection and preservation of forests, recreational services, etc. In addition to harvesting and wood production, some other criteria are receiving increased attention in choosing the ways of forest management. In other words, forests are simultaneously used for multiple purposes. Multiple benefits and many advantages provided by forests as well as the non-market nature of a part of these products, make the planning and decision making in forestry especially demanding. This has led to a need for models that can be applied in multifunctional sustainable forest management. In particular, such support, through various methods and models is needed in planning and predicting, as well as in the analysis of forest management results.

Forest management involves the decision making related to the organisation, use and conservation of forests. Management decisions are made both for long-term planning and daily activities. Good forest management requires solution of the issues related to problems of energy, raw materials and life quality. Mathematical models are not new in forestry. The multiplicity of the available data on forests requires computer-aided mathematical methods. The problems of forest management involve a variety of different variables. They may be biological, such as growth and increment, type of soil; economic, such as the price of timber and labour costs; and social, such as ecological laws. All these variables and their interrelations make up a system. The complexity of forestry systems makes predicting of consequences of the decisions made a difficult task. This is where models come in use. Most models calculate the consequences of particular decisions. The models may be classified according to their properties. Thus, they may be deterministic, or stochastic; they may optimise one, or several goals and they use a particular algorithm. First models used linear programming. Many authors used dynamic programming as a method for making a series of optimal decisions (Amidon & Atkin, 1968; Brodie & Kao, 1979; Zadnik, 1990). Realising the necessity of stochastic methods in forest management, the Markov process was introduced into forestry (Hool, 1966). Multiple uses in forest management were first expressed through goal programming (Field, 1973; Mendoza, 1986). Goal programming uses the weights that are the reflections of the significances of each criterion. To join these weights is the greatest problem in goal programming, so that different authors suggested different methods (Bare & Mendoza, 1988; Gong, 1992; Howard & Nelson, 1993). Group decision is the most complex form of decision-making. It basically does not differ from the multi-criteria decision. The only difference is organisation of hierarchy and the sequence of taking the individual steps in the decision.

Planning and decision making in forestry is especially complex because of multiple objectives of forest management, and numerous wide ranging, often hardly comparable and conflicting criteria and interests that influence the decision making process. Multi-criteria methods can thereby facilitate the decision making process and reduce the risks and challenges in today's demanding and complex forest management planning. It is sure that MCDM and operations research can not resolve all issues and problems in forestry, but they can serve as a platform on which the results of different scientific fields can be used in a comprehensive decision-making process.

It should also be pointed out that managing any organization requires the ability to effectively assess and analyze information generated in the business process. For organizations, such as forestry companies, which manage natural resources and by business decisions affect the environment, that is from the viewpoint of ecological acceptability and environmental management even more critical. Development and application of methods that have not been traditionally used in natural resources management can provide a valuable assistance at the strategic, tactical and operational level of planning and decision making. Methods that have in this respect experienced a wide range of applications in recent years are for example AHP and DEA.

This paper, besides AHP and DEA also presents the other major MCDM methods: MAUT, ouranking methods, voting methods and SMAA. Paper gives the basic features of methods and a brief overview of forestry problems and areas where they have been applied so far. The aim was to provide information on existing experience, and thus contribute to making forestry profession aware of the significance and potential role that MCDM methods can play in forestry. Many cited articles can also be a valuable reference source for students, researchers, forestry experts and practitioners. The results show that in the last 30 years a significant number of forestry MCDM papers was published dealing with various forestry issues and problems such as harvesting, biodiversity, sustainability, regional planning, etc. Frequency of published papers shows that the number of such papers is increasing at a very high rate what indicates a trend of increased use of MCDM in forestry in recent years.

In this very dynamic period of natural resources management, when forestry experts face the challenges of professional and responsible management of forests and forest land, having to observe at the same time the protection requirements of their ecological, social and economic functions, as well as challenges of profitable management of forestry companies, managers need different models for converting natural, accounting, financial and many other variables and data into useful information. This paper points to the justification and possibilities of application of MCDM in multifunctional forest management, with the emphases on conservation of biodiversity, regeneration capacity and sustainable management. Paper also shows how multi-criteria methods can be used for analyzing the choice of the best or at least satisfactory decision, and thus contribute to more reliable planning and more objective decision making in forestry. It is generally considered that MCDM methods in forestry, as well as in other business systems, can be a very strong support to management and decision making.

8. References

Amidon, E. L. & Atkin, G. S. (1968). Dynamic Programing to Determine Optimum Levels of Growing Stock. *Forest Science* 14 (3): 287-291.

Ananda, J. & Herath, G. (2003). The use of Analytic Hierarchy Process to incorporate stakeholder preferences into regional forest planning. *Forest Policy and Economics*, 5 (1): 13-26.

Ananda, J. & Herath, G. (2005). Evaluating public risk preferences in forest land-use choices using multi-attribute utility theory. *Ecolgical Economics*, 55 (3): 408-419.

Arp, P.A. & Lavigne, D.R. (1982). Planning with goal programming: a case study for multiple.use of forested land. *Forestry Chronicle*, 58 (5): 225-232.

Bahovec, V. & Neralić, L. (2001). Relative efficiency of agricultural production in county districts of Croatia. *Mathematical Communications* - Supplement 1 (2001), 1: 111–119.

Bare B. B. & Mendoza G. A. (1988). Multiple Objective Forest Land Management Planning. *European Journal of Operational Research*, 34 (1): 44-55.

Bogetoft, P.; Thorsen, B.J. & Strange, N. (2003). Efficiency and merger gains in the Denish Forestry Extension Service. *Forest Science*, 49 (4): 585-595.

Brans, J.P.; Vincke, Ph. & Mareschal, B. (1986). How to select and how to rank projects: the PROMETHEE method. *European Journal of Operational Research*, 24 (2): 228-238.

Brodie, J. D. & Kao, C. (1979). Optimizing Thining in Douglas-fir with Three-descriptor Dynamic Programming to Account for Accelerated Diameter Growth. *Forest Science*, 25 (1): 665-672.

Buongiorno, J. & Svanquist, N.H. (1982). A separable goal programming model of the Indonesian forestry sector. *Forest Ecology and Management*, 4 (1): 67-78.

Charnes, A.; Cooper, W.W. & Rhodes, E. (1978). Measuring the efficiency of decision making units. *European Juornal of Operational Research*, 2: 429-444.

Charnes, A.; Cooper, W.; Lewin, A. & Seiford, L. (1994). *Data envelopment analysis, theory, methodology and applications*. Kluwer Academic Publishers, Boston.

Cooper, W.W.; Seiford, L.M. & Tone, K. (2003). *Data Envelopment Analysis – A Comprehensive Text with Models, Applications, References and DEA-Solver Software*, Kluwer Academic Publishers, p. 1–318.

Čaklović, L; Piskač, R & Šego, V. (2001). Improvement of AHP method, *Proceedings of the 8th International Conference on Operational Research-KOI 2002*, Osijek, p 13-21.

Davosir Pongrac, D. (2006). *Efikasnost osiguravajućih društava u Republici Hrvatskoj*. Magistarski rad, Ekonomski fakultet, Zagreb, str. 1–139 + III.

Diaz-Balteiro, L. & Romero, C. (1998). Modeling timber harvest scheduling problems with multiple criteria: an application to Spain. *Forest Science*, 44 (1): 47-57.

Diaz-Balteiro, L.; Herruzo, A. C.; Martinez, M. & González-Pachón, J. (2006). An analysis of productive efficiency and innovation activity using DEA: An application to Spain's wood-based industry. *Forest Policy and Economics*, 8 (7): 762-773.

Diaz-Balteiro, L. & Romero, C. (2008). Making forestry decisions with multiple criteria – a review and an assessment. *Forest ecology and management*, 255 (8-9): 3222-3241.

Faith, O.P.; Walker, P.A.; Ive, J.R. & Belbin, L. (1996). Integrating conservation and forestry production: exploring trade-offs between biodiversity and production in regional land-use assessment. *Forest ecology and management*, 85 (1-3): 251-260

Farrell, M.J. (1957). The measurement of productive efficiency. *Journal of the Royal Statistical Society*, Series A 120 (3): 253-281.

Field, D.B. (1973). Goal programming for forest management. *Forest Science*, 19 (2): 125-135.

Galanopoulos, K.; Aggelopoulos, S.; Kamenidou, I. & Mattas, K. (2006). Assesing the effects of managerial and production practices on the efficiency of commercial pig farming. *Agricultural Systems*, 88 (2-3): 125-141.

Gilliams, S.; Raymaekers, D.; Muys, B. & Orshoven, J. (2005). Comparing multiple criteria decision methods to extend a geographical information system on afforestation. *Computers and Electronics in Agriculture*, 49 (1): 142-158.

Glass, .JC.; McKillop, D.G. & O'Rourke, G. (1999). A cost indirect evaluation of productivity change in UK universities. *Journal of Productivity Analysis* 10 (2): 153–75.

Gong, P. (1992). Multiobjective Dynamic Programming for Forest Resource Management. *Forest Ecology and Management*. 48 (1-2): 43-54.

Hallefjord, A.; Jornsten, K. & Eriksson, O. (1986). A long range forestry planning problem with multiple objectives. *European Journal of Operational Research*, 26 (1): 123-133.

Heinonen, T. & Pukkala, T. (2004). A comparison of one- and two-compartement neighbourhoods in heuristic search with spatial forest management goals. *Silva Fennica*, 38 (3): 319-332.

Herath, G. & Prato, T. (2006). *Using multi-criteria decision analysis in natural resource management*, Ashgate publishing, 239 p., Hampshire, England

Hiltunen, V.; Kangas, J. & Pykäläinen, J. (2008). Voting methods in strategic forest planning – Experiences from Metsähallitus. *Forest Policy and Economics*, 10 (3): 117-127.

Hool, J. N. (1966). A Dynamic Programming Markov Chain Approach to Forest Production Control, *Forest Science Monograph*, 12: 1-26.

Howard, A.F. & Nelson, J.D. (1993). Area-based harvest scheduling and allocation of forest land using methods for multiple-criteria decision making. *Canadian Journal of Forest Research,* 23 (2): 151-158.

Huth, A.; Drechsler, M. & Kohler, P. (2005). Using multicriteria decision analysis and a forest growth model to assess impacts of tree harvesting in Dipterocarp lowland rain forests. *Forest Ecology and Management,* 207 (1-2): 215-232.

Hyberg, B.T. (1987). Multiattribute decision theory and forest management; a discussion and application. *Forest Science,* 33 (4): 835-845.

Kahraman, C. (2008). *Fuzzy multi-criteria decision making: theory and applications with recent developments,* 591 p., Berlin/Heidelberg.

Kajanus, M.; Kangas, J. & Kurttila, M. (2004). The use of value focused thinking and the A'WOT hybrid method in tourism management. *Tourism Management,* 25 (4): 499-506.

Kangas, J. (1992). Multiple-use planning of forest resources by using analytic hierarchy process. *Scandinavian Journal of Forest Research,* 7 (1-4): 259-268.

Kangas, J. (1993). A multi-attribute preference model for evaluating the reforestation chain alternatives of a forest stand. *Forest Ecology and Management,* 59 (3-4): 271-288.

Kangas, J. (1994). An approach to public participation in strategic forest management planning. *Forest Ecology and Management,* 70 (1-3): 75-88.

Kangas, A.; Kangas, J. & Pykalainen, J. (2001). Outranking methods as tools in strategic natural resources planning. *Silva fennica,* 35 (2): 215-227.

Kangas, J.; Hokkanen, J.; Kangas, A.; Lahdelma, R. & Salminen, P. (2003). Applying stochastic multicriteria acceptability analysis to forest ecosystem management with both cardinal and ordinal criteria. *Forest Science,* 49 (6): 928-937.

Kangas, J., Kangas, A., 2004: Multicriteria approval and SMAA-O in natural resource decision analysis with both cardinal and ordinal criteria. Journal of Multi-Criteria Decision Analysis 12 (1): 3-15.

Kangas, J.; Store, R. & Kangas, A. (2005). Socioecological landscape planning approach and multicriteria acceptability analysis in the multiple-purpose forest management. *Forest Policy and Economics,* 7 (4): 603-614.

Kangas, J. & Kangas, A. (2005). Multiple criteria decision support in forest management – the approach, methods applied, and experiences gained. *Forest ecology and management,* 207 (1-2): 133-143.

Kangas, A. (2006). The risk of decision making with incomplete criteria weight information. *Canadian Journal of Forest Research,* 36 (1): 195-205.

Kangas, A.; Kangas, J.; Lahdelma, R. & Salminen, P. (2006). Using SMAA-2 method with dependent uncertainties for strategic forest planning. *Forest Policy and Economics,* 9, (2): 113-125.

Kant, S. & Lee, S. (2004). A social choice approach to sustainable forest management: an analysis of multiple forest values in Northwestern Ontario. *Forest Policy and Economics,* 6 (3-4): 215-227.

Kao, C. (1998). Measuring the efficiency of forest districts with multiple working circles. *Journal of the Operational Research Society,* 49 (6): 583-590.

Kao, C. & Brodie, J.D. (1979). Goal programming for reconciling economic, even flow and regulation objectives in forest harvest scheduling. *Canadian Journal of Forest Research*, 9 (4): 525-531.

Keeney, R.L. & Raiffa, H. (1976). *Decisions with multiple objectives: preferences and value tradeoffs*. John Wiley & Sons, NY.

Koksalan, M.M. & Zionts, S. (2001). *Multiple criteria decision making in the new millennium*. Springer, 478 p., Berlin/Heidelberg.

Krč, J. (1999). *Večkriterijalno dinamično vrednotenje tehnoloških, ekonomskih, socialnih in ekoloških vplivov na gospodarjenje z gozdovi*. Disertacija, Biotehniška fakulteta, Univerza v Ljubljani, 174 str. Ljubljana.

Kurttila, M.; Pesonen, M.; Kangas, J. & Kajanus, M. (2000). Utilizing the analytical hierarchy process (AHP) in SWOT analysis – a hybrid method and its application to a forest certification case. *Forest Policy and Economics*, 1 (1): 41-52.

Kurttila, M.; Pykalainen, J. & Leskinen, P. (2006). Defining the forest landowner's utility-loss compensative subsidy level for a biodiversity object. *European Journal of Forest Research*, 125 (1): 67-78.

Lahdelma, R., Hokkanen, J., Salminen, P., 1998: SMAA – Stochastic multiobjective acceptability analysis. European Journal of Operational Research 106 (1): 137-143.

Laukkanen, S.; Kangas, A. & Kangas, J. (2002). Applying voting theory in natural resource management: a case of multiple-criteria group decision support. *Journal of Environmental Management*, 64 (2): 127-137.

Laukkanen, S.; Palander, T.; Kangas, J., & Kangas, A. (2005). Evaluation of the multicriteria approval method for timber-harvesting group decision support. *Silva Fennica*, 39 (2): 249-264.

LeBel, L.G. (1996). *Performance and efficiency evaluation of logging contractors using Data envelopment analysis*. Dissertation, Virginia Polytechnic Institute and State University. Blacksburg, 201 p.

Leskinen, P.; Viitanen, J.; Kangas, A. & Kangas, J. (2006). Alternatives to incorporate uncertainty and risk attitude in multicriteria evaluation of forest plans. *Forest Science*, 52 (3): 304-312.

Lexer, M.J.; Honniger, K., Scheifinger, H.; Matulla, C.; Groll, N. & KrompKolb, H. (2000). The sensitivity of central European mountain forests to scenarios of climatic change: methodological frame for a large-scale risk assessment. *Silva Fennica*, 34 (2): 113-129.

Liu, A.; Collins, A. & Yao, S. (1998). A multi-objective and multi-design evaluation procedure for environmental protection forestry. *Environmental and Resource Economics*, 12 (2): 225-240.

Maness, T. & Farrell, R. (2004). A multi objective scenario evaluation model for sustainable forest management using criteria and indicators. *Canadian Journal of Forest Research*, 34 (10): 2004-2017.

Maystre, L.Y.; Pictet, J. & Simos, J. (1994). *Methodes multicriteres ELECTRE*. Presses Polytechniques et Universitaires Romandes, Lausanne, Switzerland.

Mendoza, G.A. (1986). A Heuristic Programming Approach in Estimating Efficient Target Levels in Goal Programming. *Canadian Journal of Forest Resources,* 16 (2): 363-366.

Mendoza, G.A. & Sprouse, W. (1989). Forest planning and decision making under fuzzy environments: an overview and illustrations. *Forest Science,* 35 (2): 481-502.

Mendoza, G.A. & Prabhu, R. (2003). Qualitative multi-criteria approaches to assessing indicators of sustainable forest resource management. *Forest Ecology and Management,* 174 (1-3): 329-343.

Mendoza, G.A. & Dalton, W.J. (2005). Multi-stakeholder assessment of forest sustainability: multi-criteria analysis and a case of the Ontario forest assessment system. *Forestry Chronicle,* 81 (2): 222-228.

Moro, M.; Šporčić, M.; Šegotić, K.; Pirc, A. & Ojurović, R., 2010: The multi-criteria model for optimal selection of croatian wood industry companies. *Proceedings of International scientific conference Wood processing and furniture manufacturing: present conditions, opportunities and new challanges,* Vyhne, Slovakia, 06.-08. October 2010., p. 117-123.

Murray, D.M. & Gadow, K. (1991). Prioritizing mountain catchment areas. *Journal of Environmental Management,* 32 (4): 357-366.

Nousiainen, I.; Tahvanainen, L. & Tyrvainen, L. (1998). Landscape in farm-scale land-use planning. *Scandinavian Journal of Forest Research,* 13 (1-4): 477-487.

Nyrud, A.Q. & Bergseng, E.R. (2002). Production efficiency and size in Norwegian sawmilling. *Scandinavian Journal of Forest Research,* 17: 566-575.

Pauwels, D.; Lejeune, P. & Rondeux, J. (2007). A decision support system to simulate and compare silvicultural scenarios for pure even-aged larch stands. *Annals of Forest Science,* 64 (3): 345-353.

Posavec, S., (2001). A Discussion on the Methods of Assessing Forest Values. *Šumarski list,* 125 (11-12): 611-617.

Posavec, S. (2005). *Dynamic Model of Forest Evaluation Methods,* Disertation, Faculty of Forestry University of Zagreb, 140 pages.

Posavec, S.; Šegotić, K. & Čaklović, L. (2006). Selection of biological parameters in the evaluation of natural resources. *Periodicum Biologorum,* 108 (6): 671-676.

Pukkala, T. (1998). Multiple risks in multi-objective forest planning integration and importance. *Forest Ecology and Management,* 111 (2-3): 265-284.

Pykalainen, J. Kangas, J. & Loikkanen, T. (1999). Interactive decision analysis in participatory strategic forest planning: experiences from state owned boreal forests. *Journal of Forest Economics,* 5 (3): 341-364.

Rauscher, H.M.; Lloyd, F.T.; Loftis, D.L. & Twery, M.J. (2000). A practical decision-analysis process for forest ecosystem management. *Computers and Electronics in Agriculture,* 27 (1-3): 195-226.

Rothley, K.D. (1999). Designing bioreserve networks to satisfy multiple, conflicting demands. *Ecological Applications,* 9 (3): 741-750.

Sarkis, J.; Weinrach, J. (2001). Using data envelopment analysis to evaluate environmently conscious waste treatment technology. *Journal of Cleaner Production,* 9 (5): 417-427.

Saty, T.L. (1980). *The analytical hierarchy process.* McGraw-Hill, New York.

Saty, T.L. (2001). *Decision making with dependance and feedback - the analytic network process*. RWS Publications, Pittsburgh.

Sheldon, G.M. (2003). The efficiency of public employment services. A nonparametric matching function analysis for Switzerland. *Journal of Productivity Analysis*, 20: 49-70.

Shields, D.J.; Tolwinski, B. & Kent, B.M. (1999). Models for conflict resolution in ecosystem management. *Socio-Economic Planning Sciences*, 33 (1): 61-84.

Sowlati, T. & Vahid, S. (2006). Malmquist productivity index of the manufacturing sector in Canada from 1994 to 2002, with a focus on the wood manufacturing sector. *Scandinavian Journal of Forest Research*, 21 (5): 424-433.

Šegotić, K., (2001). Better Decisions with Quantitative Analysis in Process, In: *Logističko-distribucijski sistemi*, p 240-245, Zvolen

Šegotić, K., Šporčić, M., Martinić, I., 2003: The choice of a working method in forest stand thinning. SOR '03 Proceedings – The 7th International Symposium on Operational Research in Slovenia, Podčetrtek, Slovenia, September 24-26, 2003., p. 153-159.

Šegotić, K.; Šporčić, M. & Martinić, I. (2007). Ranking of the mechanisation working units in the forestry of Croatia. *SOR '07 Proceedings of the 9th International Symposium on Operational Research*, Nova Gorica, Slovenia, September 26-28, 2007., p. 247-251.

Šporčić, M.; Šegotić, K. & Martinić, I. (2006). Efficiency of wood transport by truck assamblies determined by Data Envelopment Analysis. *Glasnik za šumske pokuse*, pos. izdanje 5: 679-691.

Šporčić, M. (2007). *Evaluation of business success of organisational units in forestry by nonparametric model*. Disertation, Faculty of Forestry University of Zagreb, 112 pages.

Šporčić, M.; Martinić, I.; Landekić, M. & Lovrić, M. (2008). Data Envelopment Analysis as the efficiency measurement tool – possibilities of application in forestry. *Nova mehanizacija šumarstva*, vol. 29: 51-59.

Šporčić, M.; Martinić, I.; Landekić, M. & Lovrić, M. (2009). Measuring efficiency of organisational units in Forestry by nonparametric model. *Croatian Journal of Forest Engineering*, vol. 30 (1): 1-13.

Šporčić, M.; Martinić, I. & Šegotić, K. (2009). Application of 'Data Envelopment Analysis' in ecological research of maintenance of forestry mechanisation. *Strojniški vestnik – Journal of Mechanical Engineering*, 55 (10): 599-608.

Šporčić, M.; Landekić, M.; Lovrić, M.; Bogdan, S. & Šegotić, K. (2010). Multiple criteria decision making in forestry – methods and experiences. *Šumarski list*, 134 (5-6): 275-286.

Tarp, P. & Helles, F. (1995). Multi-criteria decision making in forest management planing – an overview. *Journal of Forest Econonomics*, 1 (3): 273-306.

Tavares, R. (2002). *A bibliography of Data envelopment analysis (1978-2001)*, Ructor Research Report.

Triantaphyllou, E. (2000). *Multi-criteria decision making methods: a comparative study*, Kluwer, 288 p., Dordrecht, Netherlands.

Vennesland, B. (2005). Measuring rural economic development in Norway using data envelopment analysis. *Forest Policy and Economics*, 7 (1): 109-119.

Venter, S.N.; Kühn, A.L. & Harris, J. (1998). A method for the prioritization of areas experiencing microbial pollution of surface water. *Water Science and Technology*, 38, (12): 23-27.

Vincke, Ph. (1992). *Multi-criteria decision aid*. Wiley, New York.

Walker, H.D. (1985). An alternative approach to goal programming. *Canadian Journal of Forest Research*, 15 (2): 319-325.

Wolfslehner, B.; Vacik, H. 6 Lexer, M.J. (2005). Application of the analytic network process in multi-criteria analysis of sustainable forest management. *Forest Ecology and Management*, 207 (1-2): 157-170

Yin, R. (1998). DEA: a new metodology for evaluating the performance of forest products producers. *Forest Products Journal*, 48 (1): 29-34.

Zadnik Stirn, L. (1990). Adaptive Dynamic Model for Optimal Forest Management, *Forest Ecology and Management* 31 (3): 167-188.

Permissions

The contributors of this book come from diverse backgrounds, making this book a truly international effort. This book will bring forth new frontiers with its revolutionizing research information and detailed analysis of the nascent developments around the world.

We would like to thank Jorge Martín-García and Julio Javier Diez, for lending their expertise to make the book truly unique. They have played a crucial role in the development of this book. Without their invaluable contribution this book wouldn't have been possible. They have made vital efforts to compile up to date information on the varied aspects of this subject to make this book a valuable addition to the collection of many professionals and students.

This book was conceptualized with the vision of imparting up-to-date information and advanced data in this field. To ensure the same, a matchless editorial board was set up. Every individual on the board went through rigorous rounds of assessment to prove their worth. After which they invested a large part of their time researching and compiling the most relevant data for our readers. Conferences and sessions were held from time to time between the editorial board and the contributing authors to present the data in the most comprehensible form. The editorial team has worked tirelessly to provide valuable and valid information to help people across the globe.

Every chapter published in this book has been scrutinized by our experts. Their significance has been extensively debated. The topics covered herein carry significant findings which will fuel the growth of the discipline. They may even be implemented as practical applications or may be referred to as a beginning point for another development. Chapters in this book were first published by InTech; hereby published with permission under the Creative Commons Attribution License or equivalent.

The editorial board has been involved in producing this book since its inception. They have spent rigorous hours researching and exploring the diverse topics which have resulted in the successful publishing of this book. They have passed on their knowledge of decades through this book. To expedite this challenging task, the publisher supported the team at every step. A small team of assistant editors was also appointed to further simplify the editing procedure and attain best results for the readers.

Our editorial team has been hand-picked from every corner of the world. Their multi-ethnicity adds dynamic inputs to the discussions which result in innovative outcomes. These outcomes are then further discussed with the researchers and contributors who give their valuable feedback and opinion regarding the same. The feedback is then collaborated with the researches and they are edited in a comprehensive manner to aid the understanding of the subject.

Apart from the editorial board, the designing team has also invested a significant amount of their time in understanding the subject and creating the most relevant covers. They scrutinized every image to scout for the most suitable representation of the subject and create an appropriate cover for the book.

The publishing team has been involved in this book since its early stages. They were actively engaged in every process, be it collecting the data, connecting with the contributors or procuring relevant information. The team has been an ardent support to the editorial, designing and production team. Their endless efforts to recruit the best for this project, has resulted in the accomplishment of this book. They are a veteran in the field of academics and their pool of knowledge is as vast as their experience in printing. Their expertise and guidance has proved useful at every step. Their uncompromising quality standards have made this book an exceptional effort. Their encouragement from time to time has been an inspiration for everyone.

The publisher and the editorial board hope that this book will prove to be a valuable piece of knowledge for researchers, students, practitioners and scholars across the globe.

List of Contributors

G. Perez-Verdin and J.J. Navar-Chaidez
Instituto Politécnico Nacional, Mexico

Y-S. Kim
Northern Arizona University, United States of America

R. Silva-Flores
Private Consultant, Mexico

Neelam C. Poudyal and Jacek P. Siry
Warnell School of Forestry & Natural Resources, University of Georgia, Athens, GA, USA

J. M. Bowker
USDA Forest Service, Southern Research Station, Athens, GA, USA

Ilaria Goio and Geremia Gios
Department of Economics, University of Trento, Italy

Rocco Scolozzi and Alessandro Gretter
Research and Innovation Centre, Fondazione Edmund Mach –Michele all'Adige (Trento), Italy

Kathleen McGinley, Steverson Moffat and Guy Robertson
U.S. Department of Agriculture, Forest Service, USA

Frederick Cubbage and Liwei Lin
North Carolina State University Department of Forestry and Environmental Resources, USA

Max Bruciamacchie, Serge Garcia and Anne Stenger
Laboratoire d'Economie Forestière, INRA/AgroParisTech-ENGREF, France

Teresa Fonseca and Carlos Marques
Department of Forest Sciences and Landscape Architecture (CIFAP), University of Trás-os-Montes e Alto Douro, Portugal

Bernard Parresol
USDA Forest Service, Southern Research Station, USA

François de Coligny
Institut National de la Recherche Agronomique - botAnique et bioInforMatique de l'Architecture des Plantes, INRA-AMAP, France

Nikolay Strigul
Stevens Institute of Technology, USA

Manuel Francisco Marey-Pérez, Luis Franco-Vázquez and Carlos José Álvarez-López
Universidad de Santiago de Compostela, Spain

Anusha Rajkaran and Janine B. Adams
Nelson Mandela Metropolitan University, South Africa

P. F. Newton
Canadian Wood Fibre Centre, Canadian Forest Service, Natural Resources Canada, Canada

Mario Šporčić
University of Zagreb, Faculty of Forestry, Croatia

Printed in the USA
CPSIA information can be obtained
at www.ICGtesting.com
JSHW011417221024
72173JS00004B/562